Kenn DAVIS

CCD ARRAYS, CAMERAS, and DISPLAYS

Gerald C. Holst

Copublished by

JCD Publishing
2932 Cove Trail
Winter Park, FL 32789

and

SPIE OPTICAL ENGINEERING PRESS
A publication of SPIE—The International Society for Optical Engineering
Bellingham, Washington USA

Library of Congress Cataloging-in-Publication Data

Holst, Gerald C.
   CCD Arrays, cameras, and displays / Gerald C. Holst.
      p. cm.
   Includes bibliographical references and index.
   ISBN 0-9640000-2-4 (hardcover)
   1. Charge coupled devices.   I. Title
TK7871.99.C45H65   1996
621.36'7--dc20   95-47007
                                                                  CIP

Notice:
   The Publishers and Author have taken great care in preparing the information and guidelines in this book. However, the guidelines and other material given herein are of a general nature only. The Publisher and Author take no responsibility with respect to the use of the information, guidelines, and material furnished. They assume no responsibility for any damages or costs sustained while using the guidelines.

   The Publisher and Author further disclaim any liability for any errors, omissions, or inaccuracies in the information, guidelines, and material given in this book, whether attributable to inadvertence or otherwise, and for any consequence arising therefrom. The material is published with the understanding that the Publisher and Author are supplying information. They are not attempting to render engineering or other professional services. If such services are required, the assistance of an appropriate professional should be sought.

Copyright © 1996 Gerald C. Holst

All rights reserved. No part of this book may be reproduced in any form by any means without written permission from the copyright owner.

This book is dedicated to the person who
still provides constant encouragement

Elizabeth P. Holst

PREFACE

Charge-coupled devices (CCDs) were invented by Boyle and Smith in 1970. CCDs have matured to the point where they are widely affordable and have spawned an explosion of applications. Some applications include high definition television for program production, consumer camcorders, electronic still cameras, industrial machine vision cameras, optical character readers including bar code scanners and fax machines. Applications are only limited by man's imagination.

Linear and time-delay-and-integrate (TDI) sensors are appropriate for applications where the objects are continuously moving as in the lumber, paper, textile, steel, and aluminum forming industries. These sensors provide non-contact measurement during fabrication as well as information that allows analysis of surface conditions.

The CCD architecture has three basic functions: (a) charge collection, (b) charge transfer, and (c) the conversion of charge into a measurable voltage. Since most sensors operating in the visible region use CCD architecture to read the signal, they are popularly called CCD cameras. For these devices, charge generation is often considered as the initial function of the CCD.

Although silicon CCDs respond from 10^{-4} μm to 1.1 μm, only imaging applications in the visible and near infrared are considered here. The long wavelength response may be limited by the semiconductor band gap (equivalent to 1.1 μm for silicon), diffusion length (photoelectrons recombine with holes before reaching a charge well), or infrared blocking filter (limits the response to the visible region). The short wavelength response may be limited by the polysilicon electrode transmittance (opaque below 0.4 μm) or a glass protective cover (opaque below about 0.3 μm).

This book is not a design book. It highlights those features and specifications that are most often reported in data sheets. It provides the necessary information to compare and analyze camera systems. The material in this book is representative of solid state camera operation and architecture. A specific device may not have all the features listed or may have additional features not listed.

Arrays and cameras may be specified by radiometric or photometric units (Chapter 2). Emphasis is placed on the relationship between the spectral content of the lighting compared to the sensor's spectral response. CCD array operation is described in Chapter 3 and array specifications (performance parameters) are described in Chapter 4. Array specifications, capabilities, and limitations affect the overall camera specifications. The camera manufacturer cannot increase the inherent signal-to-noise ratio of the array. He strives to maintain the array specifications without introducing additional noise or adversely affecting image quality. In Chapters 5 and 6, camera operation and specifications are related to the array values.

A camera is of no use by itself. Its value is only known when an image is evaluated. Therefore, the observer is a critical component of the imaging *system*. While the camera manufacturer strives to provide the highest possible resolution, he has no control over the display used. It is up to the end-user to select an adequate display. There is no advantage to using a high quality camera if the display cannot faithfully reproduce the image. Flat panel displays are an emerging technology but cathode ray tubes (CRT) will probably dominate the display market due to low cost, high resolution, wide color gamut, and long life. Chapter 7 discusses CRT-based display performance metrics.

Staring arrays are inherently undersampled. The highest spatial frequency that can be faithfully reproduced is the Nyquist frequency (Chapter 8). Many performance parameters are evaluated at the Nyquist frequency. The selection of Nyquist frequency is for convenience. Spatial frequencies above Nyquist frequency do not disappear. They are just distorted. Spatial sampling creates ambiguity in target edge location and produces Moiré patterns when periodic targets are viewed.

Most of the chapters discuss imaging system phenomenology with minimal mathematics. Many examples are included to illustrate basic principles and back-of-the-envelope calculations. Detailed math is found in Chapters 9 and 10 (*MTF Theory* and *System MTF*).

Many image quality metrics are linked to the system's modulation transfer function (MTF) and therefore are listed in Chapter 11. Image quality, as specified in this book, does not include data compression effects, image processing, or machine vision algorithms since these are nonlinear operators. These processes can only be evaluated on a case-by-cases basis. As digital image processing capability is incorporated into cameras, they may be called sensors with built-in computers.

The military is interested in detecting, recognizing, and identifying targets at long distances, and the minimum resolvable contrast (MRC) is an appropriate figure of merit (Chapter 12). The MRC is a system performance metric that includes both resolution (via the MTF) and sensitivity measures. It is a measure of an observer's ability to detect a bar target embedded in noise.

Although this book is titled *CCD Arrays, Cameras, and Displays*, nearly all of the concepts apply to every staring array including charge injection devices (CIDs) and active pixels sensors (APS). The electronics cannot "know" what created the signal nor the spectral content associated with the signal. Similarly, without prior training, the observer will not know what sensor was used to collect the imagery. The MTFs are generic.

Emphasis is placed upon monochrome systems. Color camera design is quite complex and represents a separate book. As such, the topics on color systems are cursory. Chapter 3 introduces basic color array operation and Chapter 5 provides some aspects of color cameras. The literature is rich with visual data on which color camera operation is based. See, for example, J. Peddie, *High-Resolution Graphics Display Systems*, Chapter 2, Windcrest/McGraw-Hill, New York, NY (1993) or A. R. Robertson and J. F. Fisher "Color Vision, Representation, and Reproduction," in *Television Engineering Handbook*, K. B. Benson, ed., Chapter 2, McGraw Hill, New York, NY (1985).

This book is for the first time CCD buyer. It explains the basic operating concepts and what the specifications represent. The book is also for the experienced buyer for it provides the background material necessary to specify cameras. The electro-optical system analyst will use the MTF and MRC theory to describe performance. For the system integrator, it provides information on the entire system: scene-to-observer. For the manager, purchasing agent, and salesperson, it is a handy reference.

The author extends his deepest gratitude to all his coworkers and students who have contributed to the ideas in this book. They are too many to mention by name. The author especially thanks all those who read draft copies of the manuscript: Constantine Anagnostopoulos, Eastman Kodak; Peter Barten, Barten Consultancy; Mike Ensminger, Ball Aerospace; Chris Fritz, Texas Instruments; Paul Gallagher, EGG Reticon; Ron Hamilton, Cohu; Richard Hofmeister, Pulnix; Herb Huey, Northrop; Lindon Lewis, Ball Aerospace; Terry Lomheim, The Aerospace Corporation; Hugh Masterson, Mitre Corporation; Ishai Nir, Princeton Instruments; Harold Orlando, Northrop; Mike Pappas, Northrop; Kenneth Parulski, Eastman Kodak; Mike Roggemann, Air Force Institute of Technology; Luke Scott, NV&ESD; Mark Sartor, Xybion; and Mark Stafford, Dalsa. Although these reviewers provided valuable comments, the accuracy of the text is the sole responsibility of the author. Jeff Odhner and Steve Kraemer provided the graphic arts.

February 1996   Gerald C. Holst

TABLE of CONTENTS

1. INTRODUCTION 1
   1.1. CCDs and CIDs 2
   1.2. Imaging system applications 3
      1.2.1. Professional television and
             consumer camcorders 5
      1.2.2. Machine vision 5
      1.2.3. Scientific 6
      1.2.4. Military 7
   1.3. Configurations 7
   1.4. Image quality 11
   1.5. Summary 12
   1.6. References 14

2. RADIOMETRY and PHOTOMETRY 16
   2.1. Radiative transfer 17
      2.1.1. Planck's blackbody law 19
      2.1.2. Conservation of energy 22
      2.1.3. Camera formula 22
   2.2. Photometry 26
      2.2.1. Units 28
      2.2.2. Natural illumination levels 30
   2.3. Sources 32
      2.3.1. Calibration sources 32
      2.3.2. Real sources 33
   2.4. Normalization 35
   2.5. Responsivity 37
   2.6. References 41

3. CCD ARRAYS 42
   3.1. CCD operation 43
   3.2. Array architecture 53
      3.2.1. Linear arrays 54
      3.2.2. Full-frame arrays 55
      3.2.3. Frame transfer 58
      3.2.4. Interline transfer 60
      3.2.5. Progressive scan 64
      3.2.6. Time-delay and integration 66

x CCD ARRAYS, CAMERAS, and DISPLAYS

    3.2.7. Color filter arrays 70
  3.3. Dark current 73
  3.4. Dark pixels 76
  3.5. Anti-blooming drain 78
  3.6. Charge transfer efficiency 79
  3.7. Charge conversion (output structure) 82
  3.8. Correlated double sampling 83
  3.9. Microlenses 85
  3.10. Video chip size 86
  3.11. Defects 87
  3.12. References 88

4. ARRAY PERFORMANCE 90
  4.1. Signal 91
    4.1.1. Spectral response 92
    4.1.2. Responsivity 97
    4.1.3. Minimum signal 102
    4.1.4. Maximum signal 102
    4.1.5. Dynamic range 103
  4.2. Noise 104
    4.2.1. Shot noise 108
    4.2.2. Reset noise 109
    4.2.3. On-chip amplifier noise 110
    4.2.4. Off-chip amplifier noise 111
    4.2.5. Quantization noise 111
    4.2.6. Pattern noise 113
    4.2.7. Photon transfer 115
  4.3. Array signal-to-noise ratio 119
  4.4. References 121

5. CAMERAS 122
  5.1. Camera operation 123
  5.2. Video formats 125
    5.2.1. Video timing 127
    5.2.2. Component/composite signals 131
    5.2.3. IRE units 132
    5.2.4. Digital television 134
    5.2.5. HDTV 135
  5.3. Consumer/broadcast cameras 137
    5.3.1. The knee 138
    5.3.2. Color correction 138
    5.3.3. Gamma compensation 142

       5.3.4. "Aperture" correction 145
   5.4. Industrial/scientific cameras 146
       5.4.1. Analog-to-digital converters 148
       5.4.2. Intensified CCDs 149
   5.5. References 153

6. CAMERA PERFORMANCE 155
   6.1. Standard camera 155
       6.1.1. Signal 156
       6.1.2. Camera signal-to-noise ratio 160
       6.1.3. Noise equivalent input 161
       6.1.4. Noise equivalent reflectance 162
   6.2. Intensified CCD camera 163
       6.2.1. Signal 164
       6.2.2. ICCD noise 165
       6.2.3. ICCD SNR 166
   6.3. References 168

7. CRT-BASED DISPLAYS 169
   7.1. The observer 171
   7.2. CRT overview 173
       7.2.1. Monochrome displays 175
       7.2.2. Color displays 176
       7.2.3. HDTV 178
   7.3. Spot size 178
   7.4. Pixels 181
   7.5. Resolution 186
       7.5.1. Vertical resolution 186
       7.5.2. Theoretical horizontal resolution 187
       7.5.3. TV limiting resolution 187
       7.5.4. MTF 188
   7.6. Addressability 189
   7.7. Shades of gray 194
   7.8. Character recognition 195
   7.9. Contrast 197
   7.10. References 198

8. SAMPLING THEORY ... 199
- 8.1. Sampling theorem ... 202
- 8.2. Aliasing ... 202
- 8.3. The detector as a sampler ... 206
  - 8.3.1. Detector MTF ... 206
  - 8.3.2. Detector array nyquist frequency ... 209
  - 8.3.3. Image distortion ... 212
- 8.4. Aliasing in frame grabbers and displays ... 217
- 8.5. References ... 218

9. MTF THEORY ... 219
- 9.1. MTF definition ... 221
- 9.2. Linear filter theory ... 223
  - 9.2.1. The EO system as a linear system ... 223
  - 9.2.2. Cascading MTFs ... 225
- 9.3. Superposition applied to optical systems ... 226
- 9.4. Contrast transfer function ... 228
- 9.5. References ... 232

10. SYSTEM MTF ... 233
- 10.1. Spatial frequency ... 235
- 10.2. Optics MTF ... 238
- 10.3. Detectors ... 240
  - 10.3.1. Diffusion MTF ... 241
  - 10.3.2. Charge transfer efficiency ... 244
  - 10.3.3. TDI ... 245
  - 10.3.4. Optical anti-alias filter ... 248
- 10.4. Motion ... 251
  - 10.4.1. Linear motion ... 251
  - 10.4.2. Random motion (jitter) ... 252
- 10.5. Electronic filters ... 254
  - 10.5.1. Digital filters ... 256
  - 10.5.2. Sample-and-hold ... 259
  - 10.5.3. Post-reconstruction filter ... 260
  - 10.5.4. Boost ... 261
- 10.6. Display ... 262
- 10.7. Eye response ... 265
- 10.8. Intensified CCD ... 269
- 10.9. Sampling effects ... 270
- 10.10. References ... 272

11. IMAGE QUALITY 275
11.1. Resolution metrics 277
11.1.1. Analog resolution metrics 279
11.1.2. Sampled data systems 281
11.1.3. Shade's equivalent resolution 285
11.2. MTFA 286
11.3. Subjective quality factor 288
11.4. Square-root integral 289
11.5. References 291

12. MINIMUM RESOLVABLE CONTRAST 293
12.1. The observer 294
12.1.1. Perceived signal-to-noise ratio 295
12.1.2. Perceived resolution 299
12.1.3. The eye/brain "filter" 300
12.2. Three-dimensional noise model 301
12.3. The MRC model 306
12.4. SNR_{TH} and t_e 310
12.5. Two-dimensional MRC 313
12.6. Range predictions 314
12.6.1. Contrast transmittance 314
12.6.2. Johnson criteria 317
12.6.3. Target transfer probability function 318
12.6.4. Range prediction methodology 320
12.6.5. Sampling effects 323
12.7. References 324

APPENDIX 327

INDEX 329

SYMBOLS

SYMBOL	DEFINITION	UNITS
α_{ABS}	spectrally averaged absorption coefficient	1/cm
$\alpha_{ABS}(\lambda)$	spectral absorption coefficient	1/cm
α_{ERROR}	target displacement in TDI systems	mm
α_G	dark current factor	numeric
α_m	display aspect ratio	numeric
β_1	eye filter 1	1/electrons^2
β_m	eye filter m	1/electrons^2
γ	gamma	numeric
$\Delta\lambda$	wavelength interval	μm
$\Delta\rho$	target-background reflectance difference	numeric
Δf_e	noise equivalent bandwidth	Hz
$\Delta n_{MINIMUM}$	minimum number of photoelectrons	numeric
Δn_{pe}	target-background photoelectrons difference	numeric
Δt	time separation between two pulses	sec
ΔV	TDI velocity error	mm/sec
ϵ	charge transfer efficiency	numeric
η	quantum efficiency	numeric
$\eta_{CATHODE}$	intensifier photocathode quantum efficiency	numeric
η_{CCD}	CCD quantum efficiency	numeric
$\eta_{PHOTOCATHODE}$	intensifier photocathode quantum efficiency	numeric
η_{SCREEN}	intensifier screen quantum efficiency	numeric
λ	wavelength	μm
λ_{AVE}	average wavelength	μm
λ_{MAX}	maximum wavelength	μm
λ_{MIN}	minimum wavelength	μm
λ_O	selected wavelength	μm
λ_P	wavelength of peak response	μm
ρ	spectrally averaged reflectance	numeric
$\rho(\lambda)$	spectral reflectance	numeric
ρ_B	background reflectance	numeric
ρ_T	target reflectance	numeric
σ_{ATM}	atmospheric absorption	1/m
σ_H	component of three-dimensional noise model	rms
σ_{HV}	component of three-dimensional noise model	rms
σ_{IIT}	image intensifier tube 1/e spot size	mm
σ_{IS}	display 1/e spot size in image space	mm
σ_R	rms value of random motion	rms mm
σ_{SPOT}	display spot 1/e intensity size	mm
σ_T	component of three-dimensional noise model	rms
σ_{TH}	component of three-dimensional noise model	rms
σ_{TVH}	component of three-dimensional noise model	rms
σ_V	component of three-dimensional noise model	rms
τ_{OPTICS}	spectrally averaged optical transmittance	numeric
$\tau_{OPTICS}(\lambda)$	spectral optical transmittance	numeric

SYMBOLS xv

Symbol	Description	Units
τ_{WINDOW}	transmittance of intensifier window	numeric
$\Phi_{CATHODE}$	flux incident onto to intensifier photocathode	watts
Φ_V	Luminous flux	lumens
$A_{CATHODE}$	area of photocathode	cm²
a_L	distance target has moved on detector	mm
A_D	detector area	cm²
A_I	area of source in image plane	cm²
A_O	lens area	cm²
A_{PIXEL}	projected CCD pixel on intensifier photocathode	cm²
A_S	source area	cm²
C	sense node capacitance	farad
c	speed of light, $c = 3 \times 10^{10}$ m/sec	
c_1	first radiation constant, $c_1 = 3.7418 \times 10^4$ watt-μm⁴/cm²	
c_2	second radiation constant, $c_2 = 1.4388 \times 10^4$ μm-K	
c_3	third radiation constant, $c_3 = 1.88365 \times 10^{22}$ photons-μm³/sec-cm²	
C_O	target's inherent contrast	numeric
C_R	target's contrast at entrance aperture	numeric
CTE	charge transfer efficiency	numeric
D	observer to display distance	m
d	detector width	mm
d_{CC}	detector pitch	mm
d_{CCH}	horizontal detector pitch	mm
d_{CCV}	vertical detector pitch	mm
d_{ERROR}	TDI error	mm
d_H	horizontal detector size	mm
D_O	optical diameter	mm
d_O	selected target dimension	mm
d_T	target detail	mm
d_V	vertical detector size	mm
E_G	detector band gap	eV
F	f-number	numeric
f_{2D}	two-dimensional spatial frequency	cy/mrad
f_{3db}	one-half power frequency	Hz
f_{BOOST}	boost frequency	Hz
f_c	cutoff frequency of an ideal circuit	Hz
f_{CLOCK}	pixel clock rate	Hz
f_{DC}	detector cutoff frequency	cy/mm
f_{elec}	electrical frequency	Hz
f_{EYE}	spatial frequency at eye	cy/deg
f_i	image spatial frequency	cy/mm
fl	focal length	m
F_{MAX}	maximum frame rate	Hz
f_N	Nyquist frequency	cy/mm
f_o	selected spatial frequency	cy/mm
f_{OC}	optics cutoff frequency	cy/mm
f_{PEAK}	peak frequency of eye MTF	cy/deg
F_R	frame rate	Hz
f_{RASTER}	raster frequency	cy/mm
f_s	sampling frequency	cy/mm
f_v	electrical frequency in the video domain	Hz
f_{VN}	Nyquist frequency in the video domain	Hz

xvi CCD ARRAYS, CAMERAS, and DISPLAYS

f_{VS}	sampling frequency in the video domain	Hz
f_x	horizontal spatial frequency	cy/mm
f_y	vertical spatial frequency	cy/mm
G	on-chip amplifier gain	numeric
G_1	off-chip amplifier gain	numeric
G_{MCP}	microchannel gain	numeric
H	target height	m
h	Planck's constant, $h = 6.626 \times 10^{-34}$ W-sec^2	
h_c	target critical dimension	m
$H_{MONITOR}$	monitor height	m
HFOV	horizontal field-of-view	mrad
J_D	dark current density	amps/cm^2
k	Boltzmann's constant, $k = 1.38 \times 10^{-23}$ W-sec/K	
k_1	a constant	numeric
$K_H(f)$	horizontal summary noise factor	
K_M	luminous efficacy, $K_M = 683$ lumens/watt (photopic)	
k_{MCP}	microchannel excess noise	numeric
$K_V(f)$	vertical summary noise factor	
L_B	background luminance	lumen/m^2-sr
L_D	Depletion length	μm
L_{DIFF}	Diffusion length	μm
L_e	spectral radiant sterance	w/(cm^2-μm-sr)
L_q	spectral photon sterance	photons/(sec-cm^2-μm-sr)
L_T	target luminance	lumen/m^2-sr
m	index	numeric
M_e	spectral radiant exitance	w/(cm^2-μm)
M_{OPTICS}	Optical magnification	numeric
$M_p(\lambda)$	spectral power	watts/μm
M_q	spectral photon exitance	photons/(sec-cm^2-μm)
N	index	numeric
n	index	numeric
N_{50}	Johnson's 50% probability	cycles
$n_{CATHODE}$	number on photons incident onto intensifier photocathode	numeric
n_{DARK}	number of dark electrons	numeric
$n_{DETECTOR}$	number of photons incidance onto detector	numeric
$N_{DETECTORS}$	total number of detectors	numeric
n_e	number of electrons	numeric
n_{EYE}	number of photons incidance onto eye	numeric
N_H	number of horizontal detectors	numeric
n_{IMAGE}	number of photons incident onto image plane	numeric
n_{LENS}	number of photons incident onto lens	numeric
N_{LINE}	number of raster lines	numeric
n_{MCP}	number on photons incident on microchannel plate	numeric
n_{pe}	number of photoelectrons	numeric
n_{pe-B}	background photoelectrons	numeric
n_{pe-T}	target photoelectrons	numeric
$n_{PHOTON-CCD}$	number on photons incident onto CCD	numeric
n_{READ}	number of pixels between active array and sense node	numeric
n_{SCREEN}	number on photons incident onto intensifier screen	numeric
N_T	equivalent number of cycles on target	cycles
N_{TDI}	number of TDI elements	numeric

SYMBOLS xvii

Symbol	Description	Units
N_{TRANS}	number of charge transfers	numeric
N_{TV}	display resolution	TVL/PH
N_V	number of vertical detectors	numeric
N_{WELL}	charge well capacity	numeric
n_{WINDOW}	number of photons incident onto intensifier	numeric
P	display line spacing or pixel spacing	mm/line
q	electron charge, $q = 1.6 \times 10^{-19}$ coul	
R	range to target	m
R_1	distance from lens to source	m
R_2	distance from lens to detector	m
R_{AVE}	spectral averaged responsivity	volts/(J-cm^{-2}) or DN/(J-cm^{-2})
R_e	spectrally averaged responsivity	amps/watt
$R_e(\lambda)$	spectral responsivity	amps/watt
R_P	peak responsivity	amps/watt
$R_{PHOTOMETRIC}$	responsivity	volts/lux
R_q	CCD spectrally averaged quantum efficiency	numeric
$R_q(\lambda)$	CCD spectral quantum efficiency	numeric
R_R	raster resolution	lines/cm
R_{TVL}	horizontal display resolution	TVL/PH
R_V	responsivity	amps/lumen
$R_{VERTICAL}$	vertical display resolution	lines
S	display spot size (FWHM intensity)	mm
T	absolute temperature	Kelvin
t_{ARRAY}	time to clock out the full array	sec
T_{ASPECT}	target aspect ratio	numeric
T_{ATM}	spectrally averaged atmospheric transmittance	numeric
$T_{ATM}(\lambda)$	spectral atmospheric transmittance	numeric
t_{CLOCK}	time between pixels	sec
t_e	eye integration time	sec
T_{fo}	fiber optic bundle transmittance	numeric
T_{illum}	absolute temperature of illuminating source	K
t_{INT}	integration time	sec
$T_{IR-FILTER}$	spectrally averaged transmittance of IR filter	numeric
t_{LINE}	video active line time	sec
$T_{RELAY\ LENS}$	relay lens transmittance	numeric
U	photoresponse nonuniformity	numeric
$V(\lambda)$	photopic or scotopic eye response	numeric
V_{CAMERA}	camera output voltage	volts
VFOV	vertical field-of-view	mrad
V_{GRID}	voltage on CRT grid	volts
V_{LSB}	voltage of one least significant bit	volts
V_{MAX}	maximum signal	numeric
V_{MIN}	minimum signal	numeric
V_{NOISE}	noise voltage after on-chip amplifier	rms volts
V_{OUT}	voltage after on-chip amplifier	volts
V_R	viewing ratio	numeric
V_{RESET}	reset voltage after on-chip amplifier	volts
V_{SCENE}	video voltage before gamma corrector	volts
V_{SIGNAL}	signal voltage after on-chip amplifier	volts
V_{VIDEO}	video voltage after gamma corrector	volts

xviii *CCD ARRAYS, CAMERAS, and DISPLAYS*

W	target width	m
$W_{MONITOR}$	monitor width	m
$\langle n_1 \rangle$	noise source 1	rms electrons
$\langle n_{ADC} \rangle$	quantization noise	rms electrons
$\langle n_{CCD-DARK} \rangle$	dark current CCD noise in an ICCD	rms electrons
$\langle n_{CCD-PHOTON} \rangle$	noise before CCD in an ICCD	rms electrons
$\langle n_{CCD} \rangle$	CCD noise in an ICCD	electrons rms
$\langle n_{DARK} \rangle$	dark current shot noise	rms electrons
$\langle n_{FLOOR} \rangle$	noise floor	rms electrons
$\langle n_{FPN} \rangle$	fixed pattern noise	rms electrons
$\langle n_{MCP} \rangle$	microchannel noise	rms electrons
$\langle n_m \rangle$	noise source m	rms electrons
$\langle n_{OFF-CHIP} \rangle$	off-chip amplifier noise	rms electrons
$\langle n_{ON-CHIP} \rangle$	on-chip amplifier noise	rms electrons
$\langle n_{PATTERN} \rangle$	pattern noise	rms electrons
$\langle n_{PC-DARK} \rangle$	intensifier dark current shot noise	rms electrons
$\langle n_{PC-SHOT} \rangle$	intensifier photon shot noise	rms electrons
$\langle n_{pe} \rangle$	photon shot noise	rms electrons
$\langle n_{PRNU} \rangle$	photoresponse nonuniformity noise	rms electrons
$\langle n_{RESET} \rangle$	reset noise	rms electrons
$\langle n_{SCREEN} \rangle$	intensifier screen noise	rms electrons
$\langle n_{SHOT} \rangle$	shot noise	rms electrons
$\langle n_{SYS} \rangle$	system noise	rms electrons
$\langle V_N \rangle$	noise voltage	rms volts

CCD ARRAYS, CAMERAS, and DISPLAYS

1
INTRODUCTION

Charge-coupled devices (CCDs) were invented by Boyle and Smith[1,2] in 1970. Since then, considerable literature[3-7] has been written on CCD physics, fabrication, and operation. However, the array does not create an image by itself. It requires an optical system to image the scene onto the array's photo sensitive area. The array requires a bias and clock signals. Its output is a series of analog pulses that represent the scene intensity at a series of discrete locations.

Devices may be described functionally according to their architecture (frame transfer, interline transfer, etc.) or by application. Certain architectures lend themselves to specific applications. For example, astronomical cameras typically use full frame arrays whereas consumer video systems use interline transfer devices.

The heart of the solid state camera is the CCD array. It provides the conversion of light intensity into measurable voltage signals. With appropriate timing signals, the temporal voltage signal represents spatial light intensities. When the array output is amplified and formatted into a standard video format, a CCD camera is created.

The array specifications, while the first place to start an analysis, are only part of the overall system performance. The *system* image quality depends on all the components. Array specifications, capabilities, and limitations are the basis for the camera specifications. Camera manufacturers cannot change these. A well-designed camera will not introduce additional noise nor adversely affect image quality provided by the array.

A camera is of no use by itself. Its value is only known when an image is evaluated. The camera output may be directly displayed on a monitor, stored on video tape or disk for later viewing, or processed by a computer. The computer may be part of a machine vision system, be used to enhance the imagery, or be used to create hard copies of the imagery. If interpretation of image quality is performed by an observer, then the observer becomes a critical component of the imaging system. Consideration of human visual system attributes should probably be the starting point of the camera design. On the other hand some machine vision systems may not create a "user friendly" image. These systems are designed from a traditional approach: resolution, high signal-to-noise ratio, and ease of operation.

Effective design and analysis require an orderly integration of diverse technologies and languages associated with radiation physics, optics, solid state sensors, electronic circuitry, human interpretation of displayed imagery (human factors), computer models, and imaging processing algorithms. Each field is complex and is a separate discipline.

1.1. CCDs and CIDs

CCD refers to a semiconductor architecture in which charge is read out of storage areas. The CCD architecture has three basic functions: (a) charge collection, (b) charge transfer, and (c) the conversion of charge into a measurable voltage. The basic building block of the CCD is the metal-oxide-semiconductor (MOS) capacitor. The capacitor is called a gate. By manipulating the gate voltages, charge can be either stored or transferred. Charge generation in most devices occurs under a MOS capacitor (also called a photo gate). For some devices (notably interline transfer devices) photodiodes create the charge. After charge generation, the transfer occurs in the MOS capacitors for all devices.

Since most sensors operating in the visible region use a CCD type architecture to read the signal, they are popularly called CCD cameras. For these devices, charge generation is often considered as the initial function of the CCD. More explicitly, these cameras should be called *solid state cameras with a CCD readout*.

CCDs and detectors can be integrated either monolithically or as hybrids. Monolithic arrays combine the detector and CCD structure on a single chip. The most common detectors are sensitive in the visible region of the spectrum. They use silicon photo gates or photodiodes and are monolithic devices. CCDs have been successfully used for infrared detectors such as Schottky barrier devices that are sensitive to 1.2 μm to 5 μm radiation.

Hybrid arrays avoid some pitfalls associated with growing different materials on a single chip and provide a convenient bridge between well-developed but otherwise incompatible technologies. HgCdTe (sensitive to 8 - 12 μm radiation) is bump bonded to a CCD readout using indium as the contact and, as such, is a hybrid array.

With charge-injection devices (CID), the pixels consist of two MOS capacitors whose gates are separately connected to rows and columns. Usually the column capacitors are used to integrate charge while the row capacitors sense the charge after integration. With CID architecture, each pixel is addressable, i.e., it is a matrix-addressable device.

CID readout is accomplished by transferring the integrated charge from the column capacitors to the row capacitors. After this nondestructive signal readout, the charge moves back to the columns for more integration or is injected (discarded) back into the silicon substrate. By suspending charge injection, the user initiates "multiple frame integration" (time lapse exposure) and can view the image on a display as the optimum exposure develops. Integration may proceed for a few milliseconds up to several hours. With individual capacitors on each sensing pixel, blooming is not transported so overloads cannot propagate.

1.2. IMAGING SYSTEM APPLICATIONS

There are five broad applications: professional television broadcast, consumer camcorder, machine vision, scientific, and military. Trying to appeal to all six applications, manufacturers use words such as low-noise, high frame rate, high resolution, reduced aliasing, and high sensitivity. These words are simply adjectives with no specific meaning. They only become meaningful when compared to another (i.e., camera A has low-noise compared to camera B).

Table 1-1 lists several design categories. While the requirements vary by category, a camera may be used for a multiple of applications. For example, a consumer video camera often is adequate for many scientific experiments. A specific device may not have all the features listed or may have additional features not listed. The separation between professional broadcast, consumer video, machine vision, scientific, and military devices becomes fuzzy as technology advances.

Color cameras are used for professional television, camcorder, and film replacement systems. With machine vision systems, color cameras are used to verify the color consistency of objects such as printed labels or paint mixture color. While color may not be the primary concern, it may be necessary for image analysis when color is the only information to distinguish boundaries.

While consumers demand color camera systems, this is not true for other applications. Depending on the application, monochrome (black and white) cameras may be adequate. A monochrome camera has higher sensitivity and therefore is the camera of choice in low-light-level conditions.

Table 1-1
DESIGN GOALS

DESIGN CATEGORY	PROFESSIONAL TELEVISION	CONSUMER CAMCORDERS	MACHINE VISION	SCIENTIFIC	MILITARY
Image processing algorithms	Gamma correction	Gamma correction	Application specific	Menu-driven multiple options	Application specific
Image processing time	Real time	Real time	Application specific with emphasis on high speed operation	Real time not usually required	Real time
Resolution	Matched to video format (e.g., EIA 170)	Matched to video format (e.g., EIA 170)	For a fixed field-of-view, increased resolution is desired	High resolution	High resolution
Dynamic range	8 bit/color	8 bit/color	8 bits/color	Up to 16 bits	10 or 12 bits
Sensitivity	High contrast targets	High contrast targets (noise not necessarily a dominant design factor)	Application specific - not necessarily an issue since lighting can be controlled	Low-noise operation	Low-noise operation

1.2.1. PROFESSIONAL TELEVISION and CONSUMER CAMCORDERS

Cameras for the professional broadcast television and consumer camcorder markets are designed to operate in real time with an output that is consistent with a standard broadcast format. The resolution, in terms of array size, is matched to the bandwidth recommended in the standard. An array that provides an output of 768 horizontal by 484 vertical pixels creates a satisfactory image for conventional television.

Consumer electronics has selected 256 (8 bits) intensity levels or gray levels. Eight bits provides an acceptable image in the broadcast and camcorder industry. The largest consumer market presently is the camcorder market.

Since CCD cameras have largely replaced image vacuum tubes, the terminology associated with these tubes is also used with CCD cameras. For example, compared to image vacuum tubes, CCDs have no image burn-in, no residual imaging, and usually are not affected by microphonics.

The current accepted meaning of high definition television (HDTV) is a television *system* providing approximately twice the horizontal and vertical resolution of present NTSC, PAL, and SECAM systems (discussed in Section 5.2.1., *Video Timing*). HDTV is envisioned as a system that provides high resolution on large displays and projection screens. The goal is to have worldwide compatibility so that HDTV receivers can display NTSC, PAL, and SECAM transmitted imagery. With today's multimedia approach, any new standard must be compatible with a variety of imaging systems ranging from 35-mm film to the various motion picture formats.

The Federal Communication Commission (FCC) is expected to adopt the HDTV standard in 1996 and broadcasting should begin between 1998 and 2000. HDTV camera descriptions, standards, and receivers will appear in many journals over the next few years.

1.2.2. MACHINE VISION

In its simplest version, a machine vision system consists of a light source, camera, and computer software that rapidly analyzes digitized images with respect to location, size, flaws, and other preprogrammed data. Unlike other types of image analysis, a machine vision system also includes a mechanism that immediately reacts to images that do not conform to the parameters stored in the computer. For example, defective parts are taken off a production line conveyor belt.

Machine vision functions include location, inspection, gauging, identification, recognition, counting, and motion tracking. These systems do not necessarily need to operate at a standard frame rate. For industrial inspection, linear arrays operating in the time-delay and integration (TDI) mode can be used to measure objects moving at a high speed on a conveyor belt.

Where a multitude of cues are used for target detection, recognition, or identification, machine vision systems cannot replace the human eye. The eye processes intensity differences over 11 orders of magnitude, color differences, and textual cues. The CCD, on the other hand, can process limited data much faster than the human. Many operations can be performed faster, cheaper, and more accurately by machines than by humans. Machine vision systems can operate 24 hours a day without fatigue. They operate consistently whereas variability exists among human inspectors. Furthermore, these cameras can operate in harsh environments that may be unfriendly to humans (e.g., extreme heat, cold, or ionizing radiation).

1.2.3. SCIENTIFIC

For scientific applications, low-noise, high responsivity, large dynamic range, and high resolution are dominant considerations. To exploit a large dynamic range, scientific cameras may digitize the signal into 12, 14, or 16 bits. Array linearity and analog-to-digital converter linearity are important. Resolution is specified by the number of detector elements and scientific arrays may have 5000 x 5000 detector elements[8]. Theoretically, the array can be any size but manufacturing considerations may ultimately limit the array size.

Low-noise means low-dark current and low-readout noise. The dark current can be minimized by cooling the CCD. Long integration times can increase the signal value so that the readout noise is small compared to the photon shot noise.

Although low-light-level cameras have many applications, they tend to be used for scientific applications. There is no industry wide definition of "low-light-level" imaging system. To some, it is simply a CCD camera that can provide a usable image when the lighting conditions are less than 1 lux. To others, it refers to an intensified camera and is sometimes called a low-light-level-television (LLLTV) system. An image intensifier amplifies a low-light-level image so that it can be sensed by a CCD camera. The image-intensifier/CCD camera combination is called an intensified CCD or ICCD. The image intensifier provides tremendous amplification but also introduces additional noise.

1.2.4. MILITARY

The military is interested in detecting, recognizing, and identifying targets at long distances. This requires high resolution, low-noise sensors. Target detection is a perceptible act. A human determines if the target is present. The military uses the minimum resolvable contrast (MRC) as a figure of merit.

CCDs are popular because of their ruggedness and small size. They can easily be mounted on remotely piloted vehicles. They are replacing wet-film systems used for mapping and photo interpretation.

1.3. CONFIGURATIONS

Imaging systems for the five broad application categories may operate in a variety of configurations. The precise setup depends on the specific requirements. Figure 1-1 is representative of a closed circuit television system where the camera output is continuously displayed. The overall image quality is determined by the camera capability (which is based on the array specifications), the bandwidth of the video format used (e.g., EIA 170, NTSC, PAL, or SECAM), the display performance, and the observer. EIA 170 was formerly called RS 170 and the NTSC standard is also called RS 170A.

Figure 1-2 illustrates a generic transmission system. The transmitter and receiver must have sufficient electronic bandwidth to provide the desired image quality. For remote sensing, the data may be compressed before the link. Compression may alter the image and the effects of compression will be seen on selected imagery. However, compression effects are not objectionable in most imagery.

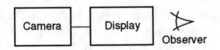

Figure 1-1. A closed circuit television system. Image quality is determined by an observer.

8 CCD ARRAYS, CAMERAS, and DISPLAYS

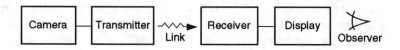

Figure 1-2. Generic remote transmission system. The transmitter and receiver must have adequate electronic bandwidth to support the image quality created by the CCD camera.

For remote applications the image may be recorded on a separate video recorder[9,10] (Figure 1-3) or may be obtained with a camcorder. The most popular player is the video cassette recorder (VCR). The recorder further modifies the image by reducing the signal-to-noise ratio (SNR) and image sharpness. As the desire for portability increases, cameras and recorders will shrink in size[10-13].

Figure 1-3. Imagery can be stored on video tape. However, the recorder circuitry may degrade the image quality. VHS is the most popular recorder.

For scientific applications, the camera output is digitized and then processed by a computer (Figure 1-4). After processing, the image may be presented on a monitor (soft copy), printed (hard copy), or stored. The digital image can also be transported to another computer via the Internet, local area net, floppy disc, or magnetic tape. For remote applications, the digital data may be stored on a digital recorder[14-16] (Figure 1-5).

Perhaps the most compelling reason for adopting digital technology is the fact that the quality of digital signals remains intact through copying and reproduction unless they are deliberately altered. Digital signal "transmission" was first introduced into tape recorders. Since a bit is either present or not, multiple generation copies retain high image quality.

INTRODUCTION 9

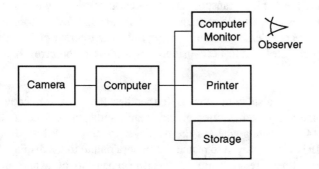

Figure 1-4. Most imagery today is enhanced through image processing. The camera may view a scene directly or may scan a document. The computer output hard copy may be used in newspapers, advertisements, or reports.

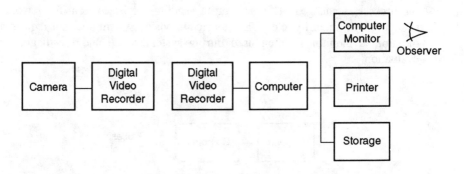

Figure 1-5. Digital systems can provide very high quality imagery. Electronic bandwidth limitations may impose data compression requirements. Data compression may alter the image but this alteration may not be obvious on general imagery.

In comparison, analog recorders, such as video home system (VHS), provide very poor quality after just a few generations. The first digital recorder used a format called D-1 and it became known as the CCIR 601 component digital standard. Digital recording formats now include[16] D-1, D-2, D-3, D-4, D-5, and D-6. Error correcting codes can be used to replace missing bits.

New digital recorders combine advanced technologies in electronics, video compression, and mechanical transport design. Real-time digital video systems operate at data rates that are faster than most computers. While errors in the imagery are never considered desirable, the eye is very tolerant of defects. In comparison, a computer would be considered worthless if the error rate was high.

When it comes to displaying images (either hard copy or soft copy) the range of the camera digitizer is often greater than the display device. A CCD may offer 12, 14, or 16 bits whereas the display may only be 8 bits. Here, a look-up-table (LUT) is employed to match the camera output to the display. This may be a simple linear relationship or specific expansion of a part of the camera's gray scale. False color can also be introduced to enhance the displayed image. False color is often useful for the human observer but serves no function in machine vision systems.

Although machine vision systems may have a monitor, they typically operate without one (Figure 1-6). That is, a computer algorithm compares an image to a stored standard. If the target does not compare favorably to the standard, then the process is changed. This may mean sending a rejected item for rework, changing light intensity, etc. While a machine vision system does not require a monitor, a monitor is often used during system set-up and for diagnostic evaluation.

Figure 1-6. Machine vision systems do not require a monitor. The computer output controls a manufacturing process. Monitors are used for set-up and diagnostic evaluation.

1.4. IMAGE QUALITY

Image quality is a subjective impression ranking of imagery from poor-to-excellent. It is a somewhat learned skill. It is a perceptual ability, accomplished by the brain, affected by and incorporating other sensory systems, emotions, learning and memory. The relationships are many and not well understood. Seeing varies between individuals and over time, in an individual. There exist large variations in an observer's judgment as to the correct rank ordering from best to worst and therefore image quality cannot be placed on an absolute scale. Visual psychophysical investigations have not measured all the properties relevant to imaging systems.

Many formulas exist for predicting image quality. Each is appropriate under a particular set of viewing conditions. These expressions are typically obtained from empirical data in which multiple observers view many images with a known amount of degradation. The observers rank-order the imagery from worst-to-best and then an equation is derived that relates the ranking scale to the amount of degradation.

Early metrics were created for film-based cameras. Image quality was related to the camera lens and film modulation transfer functions (MTFs). With the advent of televisions, image quality centered on the perception of raster lines and the minimum SNR required for *good* imagery. Here, it was assumed that the studio camera provided a perfect image and only the receiver affected the image quality.

Many tests have provided insight into image metrics that are related to image quality. Most metrics are related to the system MTF, resolution, or the signal-to-noise ratio. In general, images with higher MTFs and less noise are judged as having better image quality. There is no single *ideal* MTF shape that provides best image quality.

With a CCD camera *system*, the lens, array architecture, array noise, and display characteristics all affect *system* performance. Only an end-to-end assessment will determine the overall image quality. For example, there is no advantage to using a high-quality camera if the display cannot produce a faithful image. Often, the display is the limiting factor in terms of image quality and resolution. No matter how good the camera is, if the display resolution is poor, then the overall system resolution is poor. Only if the display's contrast and spatial resolution are better than the camera will the camera image quality be preserved.

For consumer applications, system resolution is important with SNR being secondary. Scientific applications may place equal importance on resolution and MTF. The military, interested in target detection, couples the system MTF and system noise to create the perceived signal-to-noise ratio (SNR_P):

$$SNR_P = k \frac{MTF_{SYS} \Delta I}{(system\ noise)} \frac{1}{(eye\ spatial\ filter)(eye\ temporal\ filter)} \quad (1\text{-}1)$$

ΔI is the intensity difference between the target and its immediate background. k is a proportionality constant that depends on the specific camera characteristics (aperture diameter, focal length, quantum efficiency, etc.). When the perceived SNR is above a threshold value, the target is just perceived. Selecting a threshold SNR and solving the equation for ΔI, provides the minimum resolvable contrast (MRC). When the MRC is coupled with the target description and atmospheric transmittance effects, the target range can be predicted.

1.5. SUMMARY

Electro-optical imaging system analysis is a mathematical construct that provides an optimum design through appropriate tradeoff analyses. A comprehensive model includes the target, background, the properties of the intervening atmosphere, the optical system, detector, electronics, display, and the human interpretation of the displayed information. While any of these components can be studied in detail separately, the electro-optical imaging *system* cannot. Only complete end-to-end analysis (scene-to-observer interpretation) permits system optimization.

The system MTF is the major component of system analysis. It describes how sinusoidal patterns propagate through the system. Since any target can be decomposed into a Fourier series, the MTF approach indicates how imagery will appear on the display.

Digital processing is used for image enhancement and analysis. Because the pixels are numerical values in a regular array, mathematical transforms can be applied to the array. The transform can be applied to a single pixel, group of pixels, or the entire image. Many image processing designers think of images as an array of numbers that can be manipulated with little regard to who is the final interpreter.

INTRODUCTION   13

The sensor resolution is limited by the lens focal length, pixel size, and detector center-to-center spacing (also called detector pitch or pixel pitch). In those cameras that have an analog output, the signal is digitized externally to the imaging system. The number of samples created by the external analog-to-digital converter (ADC) is simply a function of the ADC capability. Here, the sample number can be much greater than the number of detector elements. This higher number does not create more resolution. Any image processing algorithm that operates on this higher number must take into account the sensor resolution.

Image enhancement applies to those systems that have a human observer and helps the observer extract data. Some images belong to a small precious data set (e.g., remote sensing imagery from interplanetary probes). They must be processed repeatedly to extract every piece of information. Some are part of a data stream that are examined once (e.g., real-time video) and others have become popular and are used routinely as standards. These include the three-bar or four-bar test patterns, "Lena," and the baboon.

The camera cannot perfectly reproduce the scene. The array spatially samples the image and noise is injected by the array electronics. Spatial sampling creates ambiguity in target edge location and produces Moiré patterns when viewing periodic targets. While this is a concern to scientific and military applications, it typically is of little consequence to the average professional television broadcast and consumer markets.

The overall system may contain several independent sampling systems. The array spatially samples the scene, the computer may have its own digitizer, and the monitor may have a limited resolution. A monitor pixel may or may not represent a pixel in camera space. The researcher must understand the differences among the sampling lattices.

Table 1-1 (page 4) listed several design goals by application category. To some extent these goals are incompatible, thereby dictating design compromises. While the requirements vary by category, a camera may be used for a multitude of applications. For example, a consumer video camera often is adequate for many scientific experiments. The demand for machine vision systems is increasing dramatically. Smaller target detail can be discerned with magnifying optics. However, this reduces the field-of-view. For a fixed field-of-view, higher resolution (more pixels) cameras are required to discern finer detail. When selecting an imaging *system*, the environment, camera, data storage, and final image format must be considered (Table 1-2).

14 CCD ARRAYS, CAMERAS, and DISPLAYS

Table 1-2
SYSTEM DESIGN CONSIDERATIONS

ENVIRONMENT	CAMERA	TRANSMISSION and STORAGE	DISPLAY
Target size Target reflectance Distance to target Atmospheric transmittance Lighting conditions	Frame rate Detector size Array format Sensitivity Dynamic range Noise Color capability	Type of storage Storage capacity Data rate Video compression	Hard copy Soft copy Resolution

Finally, The most convincing evidence of system performance is a photograph of an image. Every time an image is transferred to another device, that device's tonal transfer and modulation transfer function will affect the displayed image. Video recorders degrade resolution. Imagery may look slightly different on different displays. Similarly, hard copies produced by different printers may appear different and there is no guarantee that the hard copy will look the same as the soft copy in every respect. Hard copies should only be considered as representative of system capability. Even with its over 100-year history, wet film developing and printing must be controlled with extreme care to create "identical" prints.

1.6. REFERENCES

1. W. S. Boyle and G. E. Smith, "Charge Coupled Semiconductor Devices," *Bell Systems Technical Journal*, Vol. 49, pp. 587-593 (1970).
2. G. F. Amelio, M. F. Tompsett, and G. E. Smith, "Experimental Verification of the Charge Coupled Concept," *Bell Systems Technical Journal*, Vol. 49, pp. 593-600 (1970).
3. M. J. Howes and D. V. Morgan, eights., *Charge-Coupled Devices and Systems*, John Walleye and Sons, NY (1979).
4. C. H. Sequin and M. F. Tompsett, *Charge Transfer Devices*, Academic Press, NY (1975).
5. E. S. Yang, Microelectronic Devices, McGraw-Hill, NY (1988).
6. E. L. Dereniak and D. G. Crowe, *Optical Radiation Detectors*, John Wiley and Sons, NY (1984).
7. A. J. P. Theuwissen, *Solid-State Imaging with Charge-Coupled Devices*, Kluwer Academic Publishers, Dordrecht, The Netherlands (1995).
8. S. G. Chamberlain, S. R. Kamasz, F. Ma, W. D. Washkurak, M. Farrier, and P.T. Jenkins, "A 26.3 Million Pixel CCD Image Sensor," in *IEEE Proceedings of the International Conference on Electron Devices*, pp. 151-155, Washington, D.C. December 10, 1995.
9. Z. Q. You and T. H. Edgar, *Video Recorders, Principles and Operation*, Prentice Hall, New York, NY (1992).
10. M. Hobbs, *Video Cameras and Camcorders*, Prentice Hall, New York, NY (1989).
11. K. Tsuneki, T. Ezaki, J. Hirai, and Y. Kubota, "Development of the High-band 8 mm Video System," *IEEE Transactions on Consumer Electronics*, Vol. 35(3), pp. 436-440 (1989).

12. M. Oku, I. Aizawa, N. Azuma, S. Okada, K. Hirose, and M. Ozawa, "High Picture Quality Technologies for and S-VHS Portable VCR," *SMPTE Journal*, Vol. 98(9), pp. 636-639 (1989).
13. T. Kawamura, S. Kasai, T. Tominaga, H. Sato, and M. Inatsu, "A New Small-Format VTR Using an 8 mm Cassette," *SMPTE Journal*, Vol. 96(5), pp. 466-472 (1987).
14. J. Watkinson, *The D-3 Digital Video Recorder*, Focal Press, Oxford, England (1992).
15. K. Suesada, K. Ishida, J. Takeuchi, I. Ogura, and P. Livingston "D-5: 1/2-in. Full Bit Rate Component VTR Format, *SMPTE Journal*, Vol. 103(8), pp. 507-516 (1994).
16. J. Hamalainen, "Video Recording Goes Digital," *IEEE Spectrum*, Vol. 32(4), pp. 76-79 (1995).

2
RADIOMETRY and PHOTOMETRY

Scenes in the visible and near infrared are illuminated by the sun, moon, starlight, night glow, or artificial sources. Since both the target and its background are illuminated by the same source, targets are detected when differences in reflectance exist. The camera's output voltage depends on the relationship between the scene spectral content and the spectral response of the camera.

Historically, cameras where designed to operate in the visible region only. For those systems, it was reasonable to specify responsivity in photometric units (e.g., volts/lux). However, when the spectral response extends past the visible region as with CCD silicon photo detectors, the use of photometric units can be confusing and even misleading.

Radiometry describes the energy or power transfer from a source to a detector. When normalized to the eye's response, photometric units are used. Radiometric and photometric quantities are differentiated by subscripts: e is used for radiometric units, q for photons, and v is used for photometric units.

The radiometric/photometric relationship between the scene and camera is of primary importance. The second radiometric/photometric consideration is the match between the display's spectral output and the eye's spectral response. Display manufacturers have considered this in detail and this reducing the burden on the system designer. However, the displayed image typically is not a precise reproduction of the scene. While the displayed image usually represents the scene in spatial detail, the intensity and color rendition may be different.

The symbols used in this book are summarized in the *Symbol List* (page xiv) and it appears after the *Table of Contents*.

2.1. RADIATIVE TRANSFER

Spectral radiant sterance, L_e, is the basic quantity from which all other radiometric quantities can be derived. It contains both the areal and solid angle concept[1] that is necessary to calculate the radiant flux incident onto a system. It is the amount of radiant flux, $\partial \Phi$, per unit wavelength radiated into a cone of incremental solid angle $\partial \Omega$ from a source whose incremental area is ∂A_S (Figure 2-1):

$$L_e = \frac{\partial^2 \Phi}{\partial A_s \, \partial \Omega} \quad \frac{watts}{cm^2 - \mu m - sr} \qquad (2\text{-}1)$$

Similarly, L_q is the spectral photon sterance expressed in photons/(sec-cm²-μm-sr). L_e and L_q are invariant for an optical system that has no absorption or reflections. That is, L_e and L_q remain constant as the radiation transverses through the optical system. Table 2-1 provides the standard radiometric units.

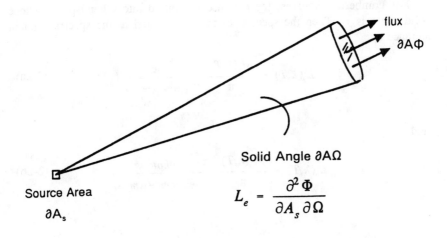

Figure 2-1. Radiant sterance.

Table 2-1
STANDARD SPECTRAL RADIOMETRIC UNITS

DEFINITION	SYMBOL	UNITS
Radiant energy	Q_e	Joule (J)
Radiant flux	Φ_e	Watts (W)
Spectral radiant intensity	I_e	W/(cm²-sr)
Spectral radiant exitance (from a source)	M_e	W/(cm²-μm)
Spectral radiant incidance (onto a target)	E_e	W/(cm²-μm)
Spectral radiant sterance	L_e	W/(cm²-μm-sr)

For Lambertian sources, the radiance is emitted into a hemisphere whose solid angle is π. Then the spectral exitance is related to the spectral radiant sterance by:

$$L_e(\lambda,T) = \frac{M_e(\lambda,T)}{\pi} \quad \frac{w}{cm^2-\mu m-sr} \quad (2\text{-}2a)$$

and

$$L_q(\lambda,T) = \frac{M_q(\lambda,T)}{\pi} \quad \frac{photons}{sec-cm^2-\mu m-sr} \quad (2\text{-}2b)$$

2.1.1. PLANCK'S BLACKBODY LAW

The spectral radiant exitance of an ideal blackbody source whose absolute temperature (color temperature) is T (Kelvin), can be described by Planck's blackbody radiation law:

$$M_e(\lambda,T) = \frac{c_1}{\lambda^5} \left(\frac{1}{e^{(c_2/\lambda T)} - 1} \right) \quad \frac{w}{cm^2-\mu m} \quad (2\text{-}3)$$

where the first radiation constant is $c_1 = 3.7418 \times 10^4$ watt-$\mu m^4/cm^2$ and the second radiation constant is $c_2 = 1.4388 \times 10^4$ μm-K. T is also called the color temperature. Figure 2-2 illustrates Planck's spectral radiant exitance in logarithmic coordinates. Since a photo detector responds linearly to the available power, linear coordinates may provide an easier representation to interpret (Figure 2-3). The curves have a maximum value when $\lambda_{MAX} T = 2898$ μm-K (Wien's displacement law).

The spectral photon exitance is simply the spectral radiant exitance divided by the energy of one photon (hc/λ):

$$M_q(\lambda,T) = \frac{c_3}{\lambda^4} \left(\frac{1}{e^{(c_2/\lambda T)} - 1} \right) \quad \frac{photons}{sec-cm^2-\mu m} \quad (2\text{-}4)$$

where the third radiation constant is $c_3 = 1.88365 \times 10^{22}$ photons-μm^3/sec-cm^2. Figures 2-4 and 2-5 provide the spectral photon exitance in logarithmic and linear coordinates respectively for sources that are typically used for CCD calibration. These curves have a maximum when $\lambda_{MAX} T = 3670$ μm-K.

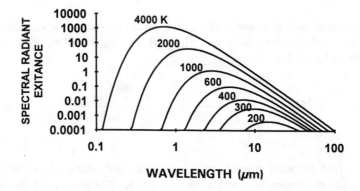

Figure 2-2. Planck's spectral radiant exitance plotted in logarithmic coordinates for T = 200, 300, ..., 4000 K. The units are watts/(cm²-μm). A source must have an absolute temperature above about 700 K to be perceived by the human eye.

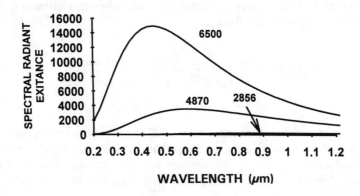

Figure 2-3. Planck's spectral radiant exitance plotted in linear coordinates for typical sources used for calibrating CCD arrays. The units are watts/(cm²-μm). The maximum values occur at 1.01, 0.60, and 0.45 μm for color temperatures of 2856, 4870, and 6500 K respectively.

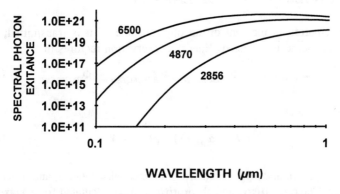

Figure 2-4. Planck's spectral photon exitance for T = 2856, 4870, and 6500 K. The maximum values occur at 1.29, 0.75, and 0.56 μm respectively. The units are photons/(sec-cm^2-μm).

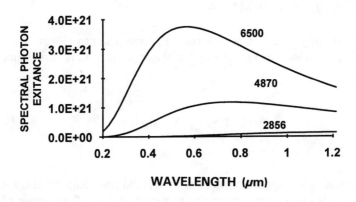

Figure 2-5. Planck's spectral photon plotted in linear coordinates. The units are photons/(sec-cm^2-μm).

2.1.2. CONSERVATION OF ENERGY

When radiation is incident on an object, some of it is transmitted, some absorbed, and some is reflected. Energy conservation requires that

$$\Phi_{TRANSMITTED} + \Phi_{ABSORBED} + \Phi_{REFLECTED} = \Phi_{INCIDENT} \quad (2\text{-}5)$$

Expressed as a ratio,

$$\tau(\lambda) + \alpha_{ABS}(\lambda) + \rho(\lambda) = 1 \quad (2\text{-}6)$$

where $\tau(\lambda)$, $\alpha_{ABS}(\lambda)$, and $\rho(\lambda)$ are the transmittance, absorptance, and reflectance respectively. The reflect*ivity*, and absorpt*ivity* can be calculated from Maxwell's equations and represent the values for an ideal material. Real materials deviate from the ideal properties and have transmitt*ance*, reflect*ance*, and absorpt*ance*.

2.1.3. CAMERA FORMULA

If an imaging system is at a distance R_1 from a source (Figure 2-6), the number of photon incident onto the optical system of area A_O is during time t_{INT} is

$$n_{LENS} = L_q \frac{A_O}{R_1^2} A_S T_{ATM} t_{INT} \quad (2\text{-}7)$$

where the small angle approximation was used (valid when $R_1 \gg D$ where $A_O = \pi D^2/4$). T_{ATM} is the intervening atmospheric transmittance. The number of on-axis photons reaching the image plane is:

$$n_{IMAGE} = L_q \frac{A_O}{R_2^2} A_S \tau_{OPTICS} T_{ATM} t_{INT} \quad (2\text{-}8)$$

τ_{OPTICS} is the system's optical transmittance.

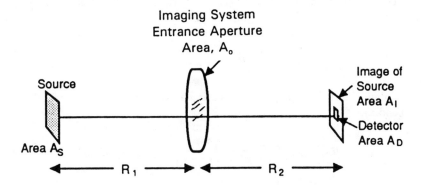

Figure 2-6. An imaging system directly viewing a source.

When the source size is much larger than the projected area of the detector ($A_I \gg A_D$), the source is said to be resolved or the system is viewing an extended source. Equivalently, the detector is flood-illuminated. In nearly all CCD applications, the source is resolved. A_D is the effective sensitive detector area. Microlenses (discussed in Section 3.9., *Microlenses*) and optical anti-alias filters (discussed in Section 10.3.4., *Optical Anti-alias Filter*) increase the effective area.

The photons incident onto the detector is simply the ratio of the areas:

$$n_{DETECTOR} = n_{IMAGE} \frac{A_D}{A_I} \qquad (2\text{-}9)$$

Using the small angle approximation for paraxial rays,

$$\frac{A_S}{R_1^2} = \frac{A_I}{R_2^2} \qquad (2\text{-}10)$$

The number of photons reaching the detector becomes:

$$n_{DETECTOR} = \frac{L_q A_O A_D}{fl^2 (1 + M_{OPTICS})^2} \tau_{OPTICS} T_{ATM} t_{INT} \qquad (2\text{-}11)$$

24 CCD ARRAYS, CAMERAS, and DISPLAYS

The optical magnification is $M_{OPTICS} = R_2/R_1$. Here, R_1 and R_2 are related to the system's effective focal length, fl, by:

$$\frac{1}{R_1} + \frac{1}{R_2} = \frac{1}{fl} \qquad (2\text{-}12)$$

Assuming a circular aperture and defining the f-number as $F = fl/D$ (see Appendix):

$$n_{DETECTOR} = \frac{\pi}{4} \frac{L_q A_D}{F^2(1+M_{OPTICS})^2} \tau_{OPTICS} T_{ATM} t_{INT} \qquad (2\text{-}13)$$

An off-axis image will have reduced incidance compared to an on-axis image by cosine$^4\theta$. As required, cosine$^4\theta$ can be added to all the equations.

All of the variables are a function of wavelength and the source photon radiant sterance depends on temperature. The number of photoelectrons generated in a solid state detector is

$$n_{pe} = \int_{\lambda_1}^{\lambda_2} R_q(\lambda) n_{DETECTOR}(\lambda) \, d\lambda \qquad (2\text{-}14)$$

or

$$n_{pe} = \int_{\lambda_1}^{\lambda_2} R_q(\lambda) \frac{\pi}{4} \frac{L_q(\lambda,T) t_{INT} A_D}{F^2(1+M_{OPTICS})^2} \tau_{OPTICS}(\lambda) T_{ATM}(\lambda) \, d\lambda \qquad (2\text{-}15)$$

where $R_q(\lambda)$ has units of electrons per photon and is simply the detector's quantum efficiency. t_{INT} becomes is the CCD integration time. As the source moves to infinity, M_{OPTICS} approaches zero.

In photography, shutter speeds (exposure times) vary approximately by a factor of two (e.g., 1/30, 1/60, 1/125, 1/250, etc.). Thus, changing the shutter speed by one setting changes n_{pe} approximately by a factor of two. f-stops have been standardized to 1, 1.4, 2, 2.8, 4, 5.6, 8..... The ratio of adjacent f-stops is $\sqrt{2}$. Changing the lens speed by one f-stop changes the f-number by a factor of $\sqrt{2}$. Here, also, the n_{pe} changes by a factor of two.

RADIOMETRY and PHOTOMETRY

In an actual application, what is of interest is the signal difference produced by a target and its immediate background. Here, both the target and the background are assumed to be illuminated by the same source (artificial lighting, sun, moon, night glow, star light). Let $\Delta\rho = \rho_T(\lambda) - \rho_B(\lambda)$ where $\rho_T(\lambda)$ and $\rho_B(\lambda)$ are the spectral reflectances of the target and background respectively. Although the number of electrons is used for CCD array calculations, the camera output is a voltage:

$$\Delta V_{camera} = G_C \Delta n_{pe} \quad (2\text{-}16)$$

where

$$\Delta n_{pe} = \int_{\lambda_1}^{\lambda_2} R_q(\lambda) \frac{\pi}{4} \frac{\Delta\rho \, L_q(\lambda, T_{ILLUM}) t_{INT} A_D}{F^2(1+M_{OPTICS})^2} \tau_{OPTICS}(\lambda) T_{ATM}(\lambda) d\lambda \quad (2\text{-}17)$$

G_C contains both the array output conversion gain (units of volts/electron) and the subsequent amplifier gain.

For back-of-the envelope calculations, the atmospheric transmittance is assumed to have no spectral features $T_{ATM}(\lambda) \approx T_{ATM}$ (discussed in Section 12.6.1., *Contrast Transmittance*). For detailed calculations, the exact spectral transmittance[2] must be used.

☞ ───

Example 2-1
VISUAL THRESHOLD

An object is heated to incandescence. What is the approximate color temperature? Let the object area be 10 cm². The observer is 1 meter from the object.

After adapting to the dark at least 60 minutes, the eye can perceive a few photons/sec (absolute threshold). Here, the pupil dilates to about 8 mm ($A_O = 5 \times 10^{-5}$ m²). When dark adapted, only the rods are functioning and the eye's spectral response is approximately from 0.38 μm to 0.66 μm (discussed in the next section). Using the radiometric equations,

$$n_{EYE} = \int_{0.38\,\mu m}^{0.66\,\mu m} \frac{A_O}{R_1^2} \frac{M_q(\lambda, T)}{\pi} A_s d\lambda \approx \frac{A_O}{R_1^2} \frac{M_q(\lambda_o, T)}{\pi} A_s \Delta\lambda \quad (2\text{-}18)$$

$\Delta\lambda = 0.28\,\mu$m. The peak scotopic eye response occurs at $\lambda_O = 0.515\,\mu$m. When the absolute temperature is 640 K, the target emits a few photons/sec. The observer can just perceive this object when he is in a completely blackened room. As the ambient illumination increases, the pupil constricts (A_O decreases). Furthermore, the eye's detection capability depends on the ambient lighting. As the light level increases, the target flux must also increase to be perceptible.

Example 2-2
REAL OBJECTS

Although not explicitly stated in Example 2-1, the calculation was performed for an ideal blackbody: $\epsilon(\lambda) = 1$. If the object was shiny metal, what is the approximate color temperature?

For metallic objects, the reflectance may be 90%. Since $\alpha(\lambda) + \rho(\lambda) = 1$, the absorptance and emittance are 10%. Thus the color temperature must increase sufficiently so that $M_q(\lambda,T)$ increases 10-fold. This occurs when the temperature is approximately 675 K. As with Example 2-1, the observer must be in a blackened room to perceive this object.

2.2. PHOTOMETRY

Photometry describes the radiative transfer from a source to a detector where the units of radiation have been normalized to the spectral sensitivity of the eye. It applies to all systems that are sensitive to visible radiation. The luminous flux emitted by a source is

$$\Phi_v = K_M \int_{0.38\mu m}^{0.75\mu m} V(\lambda)\, M_p(\lambda)\, d\lambda \quad lumens \tag{2-19}$$

where $M_p(\lambda)$ is the power in units of watts/μm. K_M is the luminous efficacy for photopic vision. It is 683 lumens/watt at the peak of the photopic curve ($\lambda \approx 0.55\,\mu$m) and 1746 lumens/watt for the scotopic region at $\lambda \approx 0.505\,\mu$m. Although both photopic and scotopic units are available, usually only the photopic units are used (Table 2-2 and Figure 2-7).

Table 2-2
PHOTOPIC AND SCOTOPIC EYE RESPONSE

Wavelength nm	Photopic $V(\lambda)$	Scotopic $V'(\lambda)$	Wavelength nm	Photopic $V(\lambda)$	Scotopic $V'(\lambda)$
380		0.00059	570	0.952	0.2076
390	0.00012	0.00221	580	0.870	0.1212
400	0.0004	0.00929	590	0.757	0.0655
410	0.0012	0.03484	600	0.631	0.03315
420	0.0040	0.0966	610	0.503	0.01593
430	0.0116	0.1998	620	0.381	0.00737
440	0.023	0.3281	630	0.265	0.00335
450	0.038	0.455	640	0.175	0.00150
460	0.060	0.567	650	0.107	0.00067
470	0.091	0.676	660	0.061	0.00031
480	0.139	0.793	670	0.032	
490	0.208	0.904	680	0.017	
500	0.323	0.982	690	0.0082	
510	0.503	0.997	700	0.0041	
520	0.710	0.935	710	0.0021	
530	0.862	0.811	720	0.00105	
540	0.954	0.650	730	0.00052	
550	0.995	0.481	740	0.00025	
560	0.995	0.3288	750	0.00012	

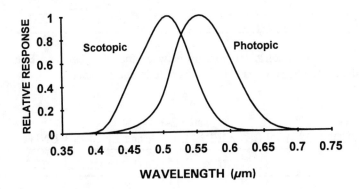

Figure 2-7. Photopic and scotopic eye responses.

2.2.1. UNITS

Unfortunately, there is overabundance of terminology being used in the field of photometry (Table 2-3). The SI units are recommended. Illumination and illuminance are sometimes used as an alternative to luminous incidence. For Lambertian sources, the luminance is emitted into a hemisphere whose solid angle is π. The luminous exitance of a Lambertian surface is simply the luminance exitance given in Table 2-3 divided by π. The terms are apostilb (abs), lambert (L), and footlambert (fL) for the SI, CGS and English systems respectively.

Figure 2-8 illustrates the geometric relationship between the SI, CGS, and English luminous incidence units. The numeric relationship is provide in Table 2-4.

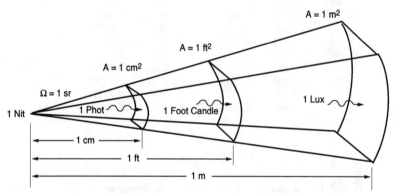

Figure 2-8. Geometric relationship between the SI, CGS, and English luminous incidence units. The solid angle is one steradian.

Table 2-4
CONVERSION BETWEEN SI, CGS, and ENGLISH UNITS

	PHOT	FOOTCANDLE	LUX
1 Phot =	1	929	1×10^4
1 Footcandle =	1.076×10^{-3}	1	10.764
1 Lux =	1×10^{-4}	0.0929	1

Table 2-3
STANDARD PHOTOMETRIC UNITS
(The SI units are recommended)

DEFINITION	SYMBOL	SI and MKS UNITS	CGS UNITS	ENGLISH UNITS
Luminous energy	Q_v	Talbot (T)	Talbot (T)	Talbot (T)
Luminous flux	Φ_v	Lumen (lm)	Lumen (lm)	Lumen (lm)
Luminous intensity	I_v	Candela (cd) lm/sr	Candela (cd) lm/sr	Candela (cd) lm/sr
Luminous exitance (from a source)	M_v	Lux (lx) lumen/m²	Phot (ph) lumen/cm²	Footcandle (fc) lumen/ft²
Luminous incidance (onto a target)	E_v	Lux (lx) Lumen/m²	Phot (ph) lumen/cm²	Footcandle (fc) lumen/ft²
Luminance (Sterance)	L_v	Nit Lumen/m²-sr or cd/m²	Stilb (sb) cd/cm²	candela/ft²

30 CCD ARRAYS, CAMERAS, and DISPLAYS

2.2.2. NATURAL ILLUMINATION LEVELS

Natural lighting levels can vary by over nine orders of magnitude (Table 2-5). The minimum level is limited by night glow and the maximum level is provided by the sun. At very low-light levels (less than 5×10^{-3} lux), the eye's rods operate (scotopic response). For light levels above 5×10^{-2} lux, the eye's cones (photopic response) respond. Between these two values, both rods and cones are operating and the eye's response is somewhere between the two values. This composite response is called the mesopic response. Table 2-6 provides typical artificial lighting levels.

The eye automatically adapts to the ambient lighting conditions to provide an optimized image. Cameras may need neutral density filters if the light level is too high. Cameras may have an automatic iris and shutter speed to optimize the image on the detector array. If the light level is too low, an intensified camera may be necessary.

Table 2-5
NATURAL ILLUMINATION LEVELS

SKY CONDITION	EYE RESPONSE	AVERAGE LUMINOUS INCIDANCE (lux)
Direct sun	Photopic	10^5
Full daylight	Photopic	10^4
Overcast sky	Photopic	10^3
Very dark day	Photopic	10^2
Twilight	Photopic	10
Deep twilight	Photopic	1.0
Full moon	Photopic	10^{-1}
Quarter moon	Mesopic	10^{-2}
Moonless, clear night (starlight)	Scotopic	10^{-3}
Moonless, overcast (night glow)	Scotopic	10^{-4}

Table 2-6
TYPICAL ARTIFICIAL ILLUMINATION LEVELS

LOCATION	LUMINOUS INCIDANCE (lux)
Hospital operating theater	10^5
TV studio	10^3
Shop windows	10^3
Drafting office	500
Business office	250
Good street lighting	20
Poor street lighting	10^{-1}

Example 2-3
JUST NOTICEABLE DIFFERENCE

An object is heated to incandescence. What is the approximate color temperature when the ambient lighting is 0.01 lux? 1000 lux? Let the object area be 1 cm^2 and $\epsilon(\lambda) = 1$.

The eye's ability to just discern intensity differences depends on the intensity of the surrounding illuminance. At 0.01 lux the object must provide 10 times more flux than the surround to be discerned by the eye. If the surround is at 700 K, than the target must be at 750 K to be perceptible. At 1000 lux, the object must provide 1000 times more flux then the surround. If the surround color temperature is 1000 K, then the target must be heated to 1350 K to be perceptible. Photometric units are not linearly related to color temperature.

These back-of-the-envelope calculations do not include the eye's spectral response (Table 2-2, page 27) and therefore only illustrate required intensity differences. The eye changes its ability to perceive differences with the surround illumination. This explains, in part, why television receiver imagery appears so bright in a darkened room and cannot be seen in bright sunlight.

2.3. SOURCES

As shown in Figure 2-3 (page 20), the peak wavelength shifts toward the blue end of the spectrum as the color temperature increases. This puts more energy in the visible region. There is approximately 70% more luminous flux available from a lamp operating at 3200 K than a light operating at 3000 K based on blackbody curves. This is the basis for using a higher color temperature lamp. However, the lifetime of tungsten halogen bulbs decreases dramatically with increasing color temperature. It is far better to use more lamps to increase the luminous flux than to increase the color temperature and sacrifice lifetime.

2.3.1. CALIBRATION SOURCES

The CIE (Commission Internationale de l'Eclairage or International Commission on Illumination) recommended four illuminants that should be used for calibration of cameras sensitive in the visible region (Table 2-7). Commercially available sources simulate these illuminants only over the visible range only (0.38 to 0.75 μm). The source intensity is not specified; only the effective color temperature is specified. Illuminants A and D_{65} are used routinely whereas illuminants B and C are of historical interest.

Illuminant A is a tungsten filament whose output follows Planck's blackbody radiation law. The other illuminants are created by placing specific filters in front of illuminant A. These filters have spectral characteristics that, when combined with illuminant A, provide the relative outputs illustrated in Figure 2-9. On a relative basis, the curves can be approximated by blackbodies whose color temperatures are 4870, 6770, and 6500 K for illuminant B, C, D_{65} respectively.

Table 2-7
CIE RECOMMENDED ILLUMINANTS

CIE ILLUMINANT	EFFECTIVE COLOR TEMPERATURE	DESCRIPTION
A	2856 K	Light from an incandescent source
B	4870 K	Average noon sunlight
C	6770 K	Average daylight (sun + sky)
D_{65} or D_{6500}	6500 K	Daylight with a corrected color temperature

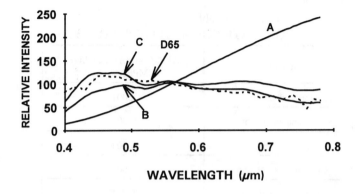

Figure 2-9. Relative output of the standard illuminants. Compare the shapes to Figure 2-3, page 20.

2.3.2. REAL SOURCES

A tungsten filament bulb emits light that is closely matched to an ideal black body radiator. This is not so for discharge lamps such as fluorescent lamps. For these lamps, a blackbody curve that approximates the output is fit to the spectral radiance (Table 2-8). This approximation is for convenience and should not be used for scientific calculations. The source spectral photon sterance is used (Equation 2-15, page 24) since the illuminating source may not be an ideal blackbody[3]. Although many discharge lamps seem uniform white in color, they have peaks in the emission spectra (Figure 2-10).

Usually there is no relationship between the color temperature and contrast of a scene and what is seen on a display. This is not the fault of the camera and display manufacturers; they strive for compatibility. The observer, who usually has no knowledge of the original scene, will adjust the display for maximum visibility and aesthetics.

The video signal is simply a voltage with no color temperature associated with it. The display electronics can be adjusted so that the display brightness appears to be somewhere between 3200 K and 10,000 K. Most displays are preset to either 6500 K or 9300 K (Figure 2-11). As the color temperature increases, whites appear to change from a yellow tinge to a blue tinge. The perceived color depends on the adapting illumination (e.g., room lighting) Setting the color temperature to 9300 K provides aesthetically pleasing imagery and this setting is unrelated to the actual scene colors.

Table 2-8
APPROXIMATE COLOR TEMPERATURE

SOURCE	APPROXIMATE COLOR TEMPERATURE
Northern sky light	7500 K
Average daylight	6500 K
Xenon (arc or flash)	6000 K
Cool fluorescent lamps	4300 K
Studio tungsten lamps	3200 K
Warm fluorescent lamps	3000 K
Floodlights	3000 K
Domestic tungsten lamps	2800 - 2900 K
Sunlight at sunset	2000 K
Candle flame	1800 K

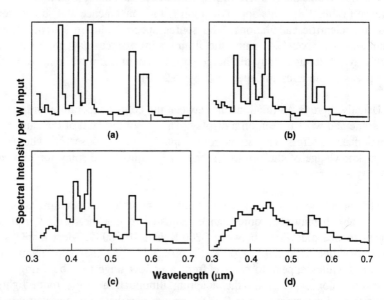

Figure 2-10. Relative output of a mercury arc lamp as a function of pressure. (a) 21 atm, (b) 75 atm, (c) 165 atm, and (d) 285 atm. (From reference 4).

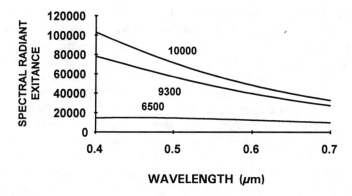

Figure 2-11. Display color temperature is typically preset at either 6500 or 9300 K. Some displays allow tuning from 3200 to 10,000 K. The displayed color temperature is independent of the scene color temperature. A scene illuminated with a 2854 K source (yellow tinge) can appear bluish when displayed at 9300 K.

2.4. NORMALIZATION

"Normalization .... is the process of reducing measurement results as nearly as possible to a common scale"[5]. Normalization is essential to insure that appropriate comparisons are made. Figure 2-12 illustrates the relationship between the spectral response of a system to two different sources. The output of a system depends on the spectral features of the input and the spectral response of the imaging system. Simply stated, the camera output depends on the source used.

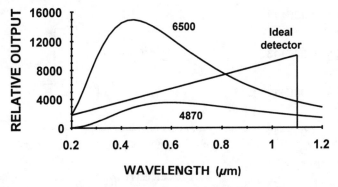

Figure 2-12. Sources with different spectral outputs can produce different system outputs. The 6500-K source provides more in-band radiant flux than the 4870-K source. The system output will be higher when viewing the 6500-K source.

36 CCD ARRAYS, CAMERAS, and DISPLAYS

Variations in output can also occur if "identical" systems have different spectral responses. Equation 2-15 (page 24) is integrated over the wavelength interval of interest. Since arrays may have different spectral responses, an imaging system whose spectral response is 0.4 to 0.7 µm may have a different responsivity than a system that operates 0.38 to 0.75 µm although both systems are labeled as visible systems. The spectral response of CCD arrays varies from manufacturer to manufacturer (discussed in Section 4.1.1., *Spectral Response*). Systems can be made to appear as equivalent or one can be made to provide better performance by simply selecting an appropriate source.

While the sensor may be calibrated with a standard illuminant, it may not be used with the same illumination. Calibration with a standard illuminant is useful for comparing camera responsivities under controlled conditions. However if the source characteristics in a particular application are significantly different from the standard illuminant, the selection of one sensor over another must be done with care. For example, street lighting from an incandescent bulb is different from that of mercury-arc lights (compare Figure 2-10 with Figure 2-12). The only way to determine the relative responsivities is to perform the calculation indicated Equation 2-14 (page 24).

The effect of ambient lighting is particularly noticeable when using low-light-level televisions at nighttime. The spectral output of the sky (no moon) and moon (clear night) are significantly different. Figure 2-13 illustrates the natural night sky irradiance. Since an abundance of photons exists in the near infrared (0.7 to 1.1 µm), most night vision systems (e.g., image intensifiers, starlight scopes, snooper scopes, etc.) are sensitive in this region.

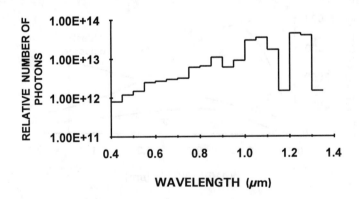

Figure 2-13. Night sky luminous incidance (From reference 6).

2.5. RESPONSIVITY

The CCD camera output depends on both the system spectral response and the color temperature of the illuminating source (Equation 2-15, page 24). When the camera spectral response is within the eye's response, then photometric units are reasonable. When the camera's response is outside the eye's response curve, photons contribute to signal but the photometry remains constant. Even though two sources may provide the same luminous incidance, the camera output can be quite different. Sometimes an IR filter is used to restrict the wavelength response. When the filter is added, the average responsivity becomes

$$R_v = \frac{\int_{\lambda_1}^{\lambda_2} T_{IR\text{-}FILTER}\, R_e(\lambda)\, M_e(\lambda,T)\, d\lambda}{683 \int_{0.38}^{0.75} V(\lambda)\, M_e(\lambda,T)\, d\lambda} \quad \frac{amps}{lumens} \quad (2\text{-}20)$$

This average type responsivity is useful for comparing the performance of cameras with similar spectral responses. It is appropriate if the source employed during actual usage is similar to the one used for the calibration.

For ideal photon detectors the spectral responsivity (in amps/watt) is:

$$R_e(\lambda) = \frac{\lambda}{\lambda_P} R_P \qquad \lambda \leq \lambda_P \quad (2\text{-}21a)$$

$$R_e(\lambda) = 0 \qquad elsewhere \quad (2\text{-}21b)$$

R_P is the peak responsivity and λ_P is the wavelength at which R_P occurs. If R_q is the quantum efficiency, $R_e = (q\lambda/hc)R_q = (\lambda/1.24)R_q$ and $R_P = R_q/E_g$. For silicon, the band gap is $E_g = 1.12$ eV resulting in a cutoff wavelength of $\lambda_P \approx 1.24/E_g \approx 1.1$ μm. q is the electron charge ($q = 1.6 \times 10^{-19}$ coul), h is Planck's constant ($h = 6.626 \times 10^{-34}$ W-sec^2), and c is the speed of light ($c = 3 \times 10^{10}$ m/sec).

38 CCD ARRAYS, CAMERAS, and DISPLAYS

The ideal silicon detector response is used to estimate the relative outputs of two systems: one with an ideal IR filter (transmittance is unity from 0.38 to 0.70 µm and zero elsewhere) and one without. The relative outputs are normalized that expected when illuminated with a CIE A illuminant (Table 2-9). The sources are considered ideal blackbodies whose illuminances follow Equation 2-4 (page 19). As the source temperature increases, the output increases but not linearly with color temperature.

Only photons within the visible region affect the number of available lumens. For the system without the IR filter, the detector is sensing photons whose wavelengths are greater than 0.7 µm even though the number of lumens is not affected by these photons (Figure 2-14). As the color temperature increases, the flux available to the observer (lumens) increases faster than the output voltage. As a result, responsivity, expressed in amps/lumen, decreases with increasing color temperature (Table 2-10). This decrease affects the full bandwidth system to a greater extent. Specifying the output in lumens is of little value for an electro-optical sensor whose response extends beyond the eye's response. Radiometric units are advised here.

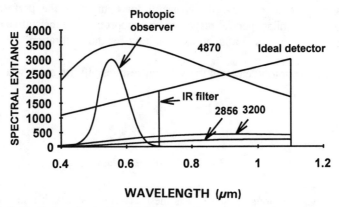

Figure 2-14. Spectral relationship between source characteristic, detector response, and the photopic observer. As the color temperature increases, the energy available to the observer increases rapidly. Therefore, the number of lumens increases dramatically.

The values in Tables 2-8 and 2-9 should be considered as illustrative. CCD detectors do not follow the ideal response (discussed in Section 4.1.1., *Spectral Response*). The IR filter cutoff also varies with by manufacturer. Only a detailed spectral response evaluation permits correct comparisons among detectors. Since there is considerable variation in spectral response and type of IR filter used, it is extremely difficult to compare systems based on responsivity values only.

Table 2-9
RELATIVE OUTPUT (Volts)
(Assuming ideal blackbodies and ideal Si spectral response)
A specific sensor may deviate significantly from these values

SOURCE TEMPERATURE	RELATIVE OUTPUT (0.38 to 1.1 μm)	RELATIVE OUTPUT (0.38 to 0.70 μm)
2856	1.00	0.196
3200	1.96	0.482
4870	15.7	6.95
6500	48.9	27.4
6770	56.5	32.4

Table 2-10
RELATIVE RESPONSIVITY (Amps/lumen)
A specific sensor may deviate significantly from these values

SOURCE TEMPERATURE	RELATIVE RESPONSIVITY (0.38 to 1.1 μm)	RELATIVE RESPONSIVITY (0.38 to 0.70 μm)
2856	1.00	0.196
3200	0.754	0.185
4870	0.387	0.171
6500	0.314	0.176
6770	0.308	0.177

Example 2-4
SOURCE SELECTION

A camera's signal-to-noise ratio is too small. The engineer can either change the light source to one with a higher color temperature or add more lamps. What are the differences?

With an ideal blackbody, the color temperature completely specifies the output luminous exitance. Increasing the color temperature (by increasing the voltage on the bulb) will increase the luminous flux. For real sources, the color temperature is used only to denote the relative spectral content. A low-wattage bulb whose output approximates a 3200-K source will have a lower photometric output than a 1000 watt bulb operating at 2856 K. Adding more lamps does not change the relative spectral content but increases the luminous flux.

Example 2-5
RELATIVE versus ABSOLUTE OUTPUT

Should a detector be calibrated with a five-watt or 1000-watt bulb?

If the relative spectral output of the two bulbs is the same, the responsivity is independent of the source flux. The bulb intensity should be sufficiently high to produce a good signal-to-noise ratio. It should not be so high that the detector saturates.

System output does not infer anything about the source other than that of an equivalent blackbody of a certain color temperature would provide the same output. This is true no matter what output units are used (volts, amps or any other arbitrary unit). These units, by themselves, are not very meaningful for system-to-system comparison. For example, ΔV_{CAMERA} (and equivalently the camera responsivity) can be increased by increasing the system gain G. As such, it is dangerous to compare system response based on only a few numbers.

Arrays may be specified by their radiometric responsivity (volts/(J-cm^{-2}) or by their photometric responsivity (volts/lux). The relationship between the two depends on the spectral content of the source and the spectral response of the array. In principle, the source can be standardized (e.g., selection of the CIE A illuminant), but the array spectral response varies among manufacturers and possibly with one manufacture if different processes are used. Therefore, there is no simple (universal) relationship between radiometric and photometric responsivities.

The correct analysis method is to perform a radiometric calculation using the source spectral output (which may not be an ideal blackbody) and using the spectral response of the camera system. Using standard sources for quality control on a production line is valid since the spectral response of the detectors should not change very much from unit-to-unit. The camera spectral response must include the optical spectral transmittance.

A 3200-K color temperature source is not well matched to the eye's response. The eye peaks at 0.555 μm but the source peaks at 0.90 μm. With silicon detectors, the spectral response tends to match the output of a 3200-K source. Thus CCD detectors (without an eye matching filter) will perform rather well with ambient lighting. For machine vision systems, the filter is not typically used. However, since the human observer is not familiar with infrared images, the imagery may appear slightly different compared to that obtained with the filter.

Low-light-level televisions can have a variety of spectral responses (discussed in Section 5.4.2., *Intensified CCDs*). The methodology presented in this chapter can be used to calculate responsivity and output voltage. The results given in Tables 2-8 and 2-9 do not apply to ICCD systems.

2.6. REFERENCES

1. C. L. Wyatt, *Radiometric System Design*, Chapter 3, Macmillan Publishing Co. New York, NY (1987).
2. The ONTAR Corporation, 129 University Road, Brookline, MA 02146-4532, offers a variety of atmospheric transmittance codes.
3. D. Kryskowski and G. H. Suits, "Natural Sources," in *Sources of Radiation*, G. J. Zissis, ed., pp. 151-209. This is Volume 1 of *The Infrared and Electro-Optical Systems Handbook*, J. S. Accetta and D. L. Shumaker, eds., copublished by Environmental Research Institute of Michigan, Ann Arbor, MI and SPIE Press, Bellingham, WA (1993).
4. E. B. Noel, "Radiation from High Pressure Mercury Arcs," *Illumination Engineering*, Vol. 36, pg. 243 (1941).
5. F. E. Nicodemus, "Normalization in Radiometry," *Applied Optics*, Vol. 12(12), pp. 2960-2973 (1973).
6. *Burle Electro-Optics Handbook*, TP-135, pg. 73, Burle Industries, Inc., Lancaster, PA (1974). Formerly known as the *RCA Electro-Optics Handbook*.

3

CCD ARRAYS

CCD (charge coupled device) refers to a semiconductor architecture in which charge is transferred through storage areas. Since most sensors operating in the visible region use a CCD architecture to move a charge packet, they are popularly called CCD arrays. The CCD architecture has three basic functions: (a) charge collection, (b) charge transfer, and (c) the conversion of charge into a measurable voltage. Since arrays operating in the visible are monolithic devices, charge generation is often considered as the initial function of the CCD. The charge is created at a pixel site in proportion to the incident light level present. The aggregate effect of all the pixels is to produce a spatially sampled representation of the continuous scene.

The basic building block of the CCD is the metal-oxide-semiconductor (MOS) capacitor. The capacitor is called a gate. Charge packets are sequentially transferred from capacitor to capacitor until they are measured on the sense node. Charge generation in most devices occurs under a MOS capacitor (also called a photo gate). For some devices (notably interline transfer devices) photodiodes create the charge. After charge generation, the transfer occurs in the MOS capacitors for all devices.

With silicon photodiodes arrays, each absorbed photon creates an electron-hole pair. Either the electrons or holes can be stored and transferred. In the following discussion, we assume an architecture where electrons are collected and transferred. The identical process exists for CCD arrays when holes are collected and stored.

Considerable literature[1-7] has been written on the physics, fabrication, and operation of CCDs. The charge transfer physics is essentially the same for all CCD arrays. However, the number of phases, and number and location of the serial shift readout registers vary by manufacturer. The description that follows should be considered as illustrative. A particular design may vary significantly from the simplified diagrams shown in this chapter. A specific device may not have all the features listed or may have additional features not listed.

Although CCD arrays are common place, the fabrication is quite complex. Theuwissen[8] provides an excellent step-by-step procedure for fabricating CCDs.

He provides a 29-step procedure supplemented with detailed diagrams. Actual CCDs may require over 150 different operations. The complexity depends upon the array architecture.

Devices may be described functionally according to their architecture (frame transfer, interline transfer, etc.) or by application. To minimize cost, array complexity, and electronic processing, the architecture is typically designed for a specific application. For example, astronomical cameras typically use full frame arrays whereas video systems generally use interline transfer devices. The separation between professional television, consumer camcorders, machine vision, scientific, and military devices becomes fuzzy as technology advances.

The symbols used in this book are summarized in the *Symbol List* (page xiv) and it appears after the *Table of Contents*.

3.1. CCD OPERATION

The basic building block of the CCD is the metal-oxide-semiconductor (MOS) capacitor. Applying a positive voltage to the gate causes the mobile positive holes in the p-type silicon to migrate toward the ground electrode since like charges repel. This region, which is void of positive charge, is the depletion region (Figure 3-1). If a photon whose energy is greater than the energy gap is absorbed in the depletion region, it produces an electron-hole pair. The electron stays within the depletion region whereas the hole moves to the ground electrode. The amount of negative charge (electrons) that can be collected is proportional to the applied voltage, oxide thickness, and gate electrode area. The total number of electrons that can be stored is called the well capacity.

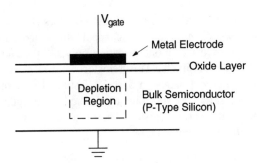

Figure 3-1. Metal-oxide-semiconductor (MOS) gate for p-type silicon. Although the depletion region (charge well) is shown to have an abrupt transition, its actual shape is gradual. That is, the well is formed by a two-dimensional voltage gradient.

As the wavelength increases, photons are absorbed at increasing depths. Very long wavelength photons may pass though the CCD and never be absorbed. A photoelectron generated deep within the substrate will experience a three-dimensional random walk until it recombines or reaches the edge of a depletion region where the electric field exists. If the diffusion length is zero, then the electron-hole pairs generated in the substrate recombine immediately. This limits the long wavelength response. If the diffusion length is extremely large, then all electrons will eventually migrate to a charge well. Doping alters the diffusion length.

The CCD register consists of a series of gates. Manipulation of the gate voltage in a systematic and sequential manner transfers the electrons from one gate to the next in a conveyor-belt-like fashion. For charge transfer, the depletion regions must overlap (Figure 3-2a). The depletion regions are actually gradients and the gradients must overlap for charge transfer to occur.

Each gate has its own control voltage that is varied as a function of time. The voltage is a square wave and is called the clock or clocking signal. When the gate voltage is low, it acts as a barrier whereas when the voltage is high, charge can be stored. Initially, a voltage is applied to gate 1 and photoelectrons are collected in well 1 (Figure 3-2b). When a voltage is applied to gate 2, electrons move to well 2 in a waterfall manner (Figure 3-2c). This process is rapid and the charge quickly equilibrates in the two wells (Figure 3-2d). As the voltage is reduced on gate 1, the well potential decreases and electrons again flow in a waterfall manner into well 2 (Figure 3-2e). Finally, when gate 1 voltage reaches zero, all the electrons are in well 2 (Figure 3-2f). This process is repeated many times until the charge is transferred through the shift register.

The CCD array is a series of column registers (Figure 3-3). The charge is kept within rows or columns by channel stops or channel blocks and the depletion regions overlap in one direction only. At the end of each column is a horizontal register of pixels. This register collects a line at a time and then transports the charge packets in a serial fashion to an output amplifier. The entire horizontal serial register must be clocked out to the sense node before the next line enters the serial register. Therefore separate vertical and horizontal clocks are required for all arrays.

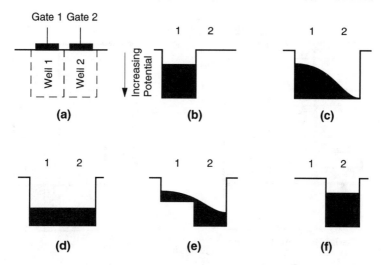

Figure 3-2. Charge transfer between two wells. (a) Adjacent wells. (b) Charge in well 1. (c) After a voltage is applied to gate 2, electrons flow into well 2. (d) Equilibration of charge. (e) Reduction of gate 1 voltage causes the electrons in well 1 electrons to fully transfer into well 2. (f) All electrons have been transferred to well 2.

Interaction between many thousands of transfers reduces the output signal. The ability to transfer charge is specified by the charge transfer efficiency (CTE). Although any number of transfer sites (gates) per detector area can be used, it generally varies from two to four.

With a four-phase device, the charge is stored under two or three wells depending upon the clock cycle. Figure 3-4 represents the steady state condition after the gates have switched and the charge has equilibrated within the wells. It takes a finite time for this to happen. Figure 3-5 illustrates the voltage levels that created the wells shown in Figure 3-4. The time that the charge has equilibrated is also shown. The clock rate is $f_{CLOCK} = 1/t_{CLOCK}$ and is the same for all phases. The output video is valid once per clock pulse (selected as t_2) and occurs after the charge has equilibrated (e.g., Figure 3-2b or Figure 3-2f). Only one master clock is required to drive the array. Its phase must be varied for the gates to operate sequentially. An anti-parallel clocking scheme is also possible. Here V_3 is the inverse of V_1 and V_4 is the inverse of V_2. That is, when V_1 is high, V_3 is low. The anti-parallel clock system is easier to implement since phasing is not required. With the anti-parallel scheme, charge is always stored in two wells. For equal pixel sizes, four-phase devices offer the largest well capacity compared to the two- or three-phase systems. With the four-phase system, 50% of the pixel area is available for the well.

46 CCD ARRAYS, CAMERAS, and DISPLAYS

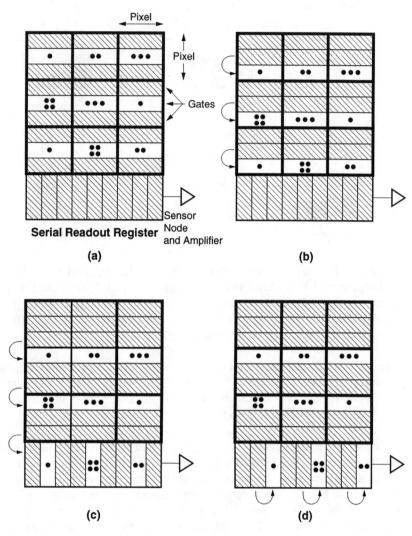

Figure 3-3. Representative CCD operation of a 3 x 3 array with three gates per pixel. Nearly all the photoelectrons generated within a pixel are stored in the well beneath that pixel. (a) The CCD is exposed to light and an electronic image is created. (b) The columns are shifted down in parallel, one gate at a time. (c) Once in the serial row register, the pixels are shifted right to the output node. (d) The entire serial register must be clocked out before the next packet can be transferred into the serial readout register.

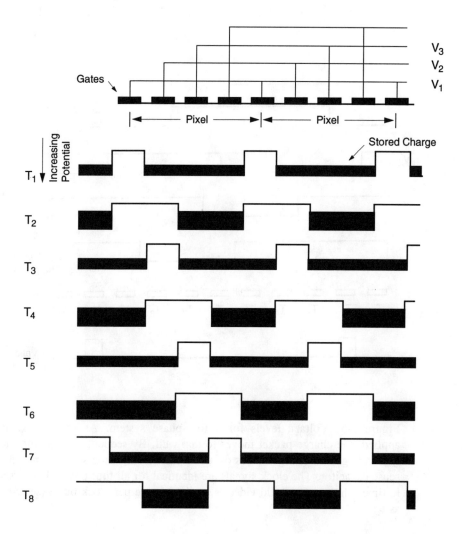

Figure 3-4. Charge transfer in a four-phase device after equilibration. This represents one column. Rows go into the paper. Charge is moved through storage sites by changing the gate voltages as illustrated in Figure 3-2. A four-phase array requires four separate clock signals to move a charge packet to the next pixel. A four-phase device requires eight steps move to move the charge from one pixel to the next.

48 CCD ARRAYS, CAMERAS, and DISPLAYS

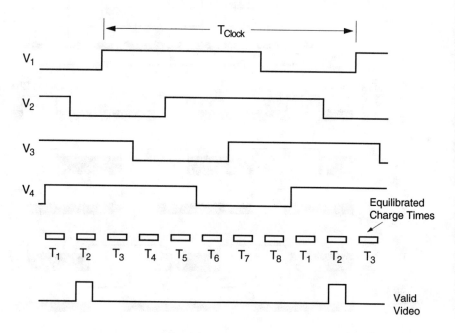

Figure 3-5. Voltage levels for a four-phase system. As the voltage is applied, the charge packet moves to that well. By sequentially varying the gate voltage, the charge moves off the horizontal shift register and onto the sense capacitor. The clock signals are identical for all four phases but offset in time (phase). The valid video is available once per clock pulse (selected as t_2).

The three-phase system (Figure 3-6 and Figure 3-7) stores the charge under one or two gates. Only 33% of the pixel area is available for the well capacity. With equal potential wells, a minimum number of three phases are required to efficiently clock out charge packets.

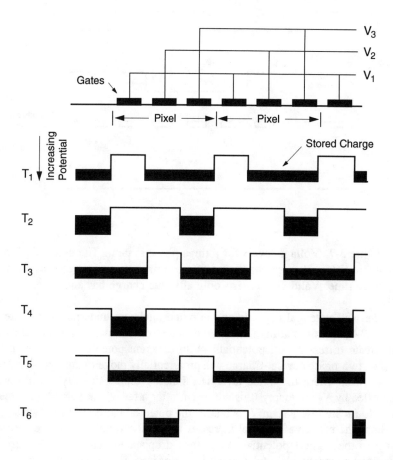

Figure 3-6. Charge transfer in a three-phase device. This represents one column. Rows go into the paper. Six steps are required to move charge one pixel.

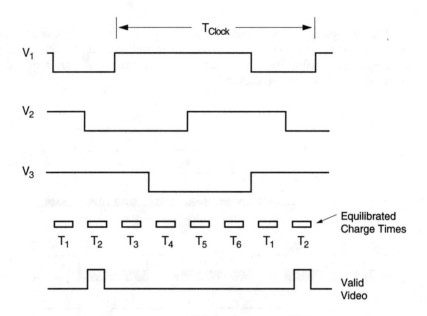

Figure 3-7. Voltage levels for a three-phase system. The clock signals are identical for all three phases but offset in time. t_2 was selected as the valid video time. Valid video exists only after the charge has equilibrated.

The well potential depends on the oxide thickness underneath the gate. In Figure 3-8, the alternating gates have different oxide thicknesses and therefore will create different well potentials. With different potentials it is possible to create a two-phase device. Charge will preferentially move to the right-hand side of the pixel where the oxide layer is thinner. Assume initially that the wells controlled by V_2 are empty (this will occur after a few clock pulses). At time t_1, both clocks are low. When V_2 is raised, the potential increases as shown at time t_2. Since the effective potential increases across the gates, the charge cascades down to the highest potential. V_2 is then dropped to zero and the charge is contained under the V_2 gate at time t_3. This process is repeated until the charge is clocked off the array. The voltage timing is shown in Figure 3-9. While Figure 3-8 illustrates variations in the oxide layer, the potential wells can also be manipulated though ion implantation. As with the four-phase device, the two-phase device can also be operated in the anti-parallel mode: V_2 is a mirror image of V_1. That is, when V_1 is high, V_2 is low.

CCD ARRAYS 51

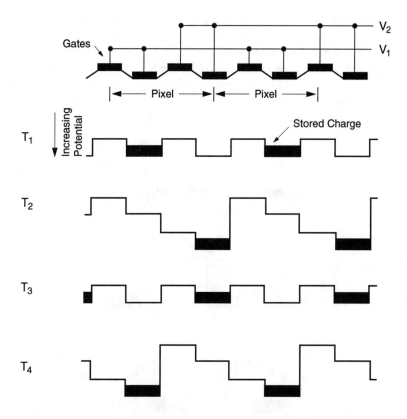

Figure 3-8. Charge transfer in a two-phase device. Well potential depends on the oxide thickness.

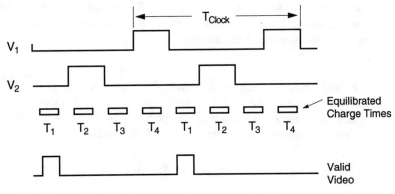

Figure 3-9. Voltage levels for a two-phase system.

52 CCD ARRAYS, CAMERAS, and DISPLAYS

The virtual phase device requires only one clock (Figure 3-10). Additional charge wells are created by internally biased gates. Charge is stored either beneath the active gate or the virtual gate. When V_1 is low, the charge will cascade down to the highest potential (which is beneath the virtual well). When V_1 is applied at t_2, the active gate potential increases and the charge move to the active gate well.

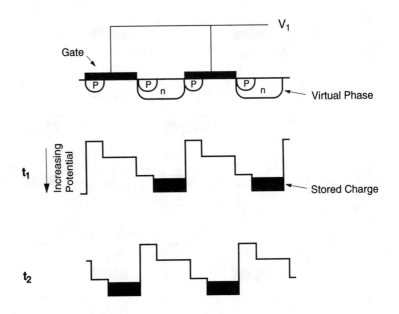

Figure 3-10. Charge transfer in a device with a virtual phase. The virtual well is created by p- and n- material implants. These ions create a fixed bias and therefore a well with fixed potential. By changing V_1, the active gate potential can be lower than the virtual well (t_1) or higher (t_2). This represents one column. Rows go into the paper.

The final operating step is to convert the charge packet to a measurable voltage. It is accomplished by a floating diode or floating diffusion. The diode, acting as a capacitor creates a voltage proportional to the number of electrons, n_e. The output voltage is sensed by a source follower and is:

$$V_{SIGNAL} = n_e \frac{Gq}{C} \qquad (3\text{-}1)$$

The gain, G, of a source follower amplifier is approximately one. q is the electronic charge and is equal to 1.6×10^{-19} coul. The signal is then amplified, processed, and digitally encoded by electronics external to the CCD sensor.

With many arrays, it is possible to shift more than one row of charge into the serial register. Similarly, it is possible to shift more than one serial register element into a summing gate just before the output node. This is called charge grouping, binning, super pixeling, or charge aggregation. Binning increases signal output and dynamic range at the expense of spatial resolution. Because it increases the signal-to-noise ratio, binning is useful for low-light-level applications in those cases where resolution is less important. Serial registers and the output node require larger capacity charge wells for binning operation. If the output capacitor is not reset after every pixel, then it can accumulate charge.

3.2. ARRAY ARCHITECTURE

Array architecture is driven by the application. Full frame and frame transfer devices tend to be used for scientific applications. Interline transfer devices are used in consumer camcorders and profession television systems. Linear arrays, progressive scan and time-delay and integration (TDI) are used for industrial applications. Despite an ever increasing demand for color cameras, black and white cameras are widely used for many scientific and industrial applications.

Progressive scan simply means the noninterlaced or sequential line-by-line scanning of the image. This is important to machine vision because it supplies accurate timing and has a simple format. Any application that requires digitization and a computer interface will probably perform better with progressively scanned imagery. However, few monitors can directly display progressive scan imagery so an interface is required. Frame capture boards provide the interface for computers.

Scientific grade arrays may be as large[9] as 5120×5120 elements (26.2×10^6 elements). While large format arrays offer the highest resolution, their use is hampered by readout rate limitations. For example, consumer camcorder systems operating at 30 frames/sec have a pixel data rate of about 10 Mpixels/sec. An array with 5120×5120 elements operating at 30 frames/sec has a pixel data rate of about 768 Mpixels/sec. Large arrays can reduce sub-array readout rates by having multiple parallel ports servicing sub-arrays. Each sub-array requires separate vertical and horizontal clock signals. The tradeoff is frame rate (speed) versus number of parallel ports (complexity of CCD design).

and interfacing with downstream electronics. Since each sub-array is serviced by different on-chip and off-chip amplifiers, the displayed image of the sub-arrays may vary in contrast and level. This is due to differences in amplifier gains and level adjustments.

"Resolution" is often specified by the number of pixels in the array. There is a perception that "bigger is better" both in terms of array size and dynamic range. Arrays may reach 9216 x 9216 with a dynamic range of 16 bits. This array requires (9216)(9216)(16) or 1.36 Gbits of storage for each image. Image compression schemes may be required if storage space is limited. The user of these arrays must decide which images are significant and through data reduction algorithms, store only those that have value. Otherwise, he will be overwhelmed with mountains of data.

3.2.1. LINEAR ARRAYS

The simplest arrangement is the linear array or single line of detectors (either photodiodes or photogates). Linear arrays are used in applications where either the camera or object is moving in a direction perpendicular to the row of sensors. They are used where rigid control is maintained on object motion such as in document scanning.

Located next to each sensor is the CCD shift register (Figure 3-11a). The shift register is also light sensitive and is covered with a metal light shield. Overall pixel size is limited by the gate size. For example, with a three-phase system, the pixel width is three times the width of a single gate. For a fixed gate size, the active pixel width can be reduced with a bilinear readout (Figure 3-11b). Since resolution is inversely related to the detector-to-detector spacing (pixel pitch), the bilinear readout has twice the resolution. For fixed pixel size, the bilinear readout increases the effective charge transfer efficiency (discussed in Section 3.6., *Charge Transfer Efficiency*).

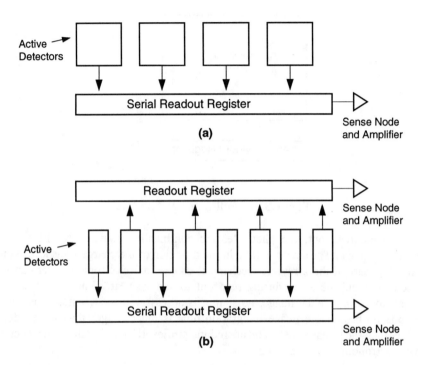

Figure 3-11. Linear array. (a) Simple structure and (b) bilinear readout.

3.2.2. FULL-FRAME ARRAYS

Figure 3-12 illustrates a full-frame transfer (FFT) array. After integration, the image pixels are read out line-by-line through a serial register that then clocks its contents onto the output sense node (see Figure 3-3, page 46). All charge must be clocked out of the serial register before the next line can be transferred. In full frame arrays, the number of pixels is often based upon powers of two (e.g., 512 x 512, 1024 x 1024) to simplify memory mapping. Scientific arrays have square pixels and this simplifies image processing algorithms.

56 CCD ARRAYS, CAMERAS, and DISPLAYS

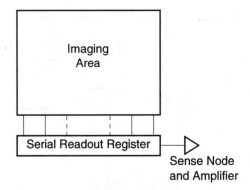

Figure 3-12. Full frame architecture.

During read out, the photosites are continually irradiated resulting in a smeared image. The smear will be in the direction of the charge transport in the imaging part of the array. A mechanical or external electronic shutter can be used to shield the array during read out to avoid smear. When using strobe lights to illuminate the image, no shutter is necessary if the transfer is between strobe flashes. If the image integration time is much longer than the readout time, then the smear may be considered insignificant. This situation often occurs with astronomical observations.

Data rates are limited by the amplifier bandwidth and the conversion capability of the analog-to-digital converter. To increase the effective readout rate, the array can be divided into sub-arrays that are read out simultaneously. In Figure 3-13, the array is divided into four sub-arrays. Since they are all read out simultaneously, the effective clock rate increases by a factor of four. Software then reconstructs the original image. This is done in a video processor external to the CCD device where the serial data is decoded and reformatted.

Large area devices often allow the user to select one sub-array for read out. Thus, the user can trade off frame rate against image size. This allows the user to obtain high frame rates over the area of interest (sub-frame). Let the active array size be m x n pixels. Let there be an additional n_{READ} pixels before the sense node. Suppose the clock operates at f_{CLOCK} pixels/sec The minimum time to clock out the array is

$$t_{ARRAY} = \left[\frac{(m + n_{READ})n}{f_{CLOCK}} \right] \quad (3\text{-}2)$$

and the maximum frame rate is

$$F_{max} = \frac{1}{t_{ARRAY}} \quad (3\text{-}3)$$

With sub-arrays, the array size is reduced and the frame rate increases. The maximum frame rate is limited by n_{READ}.

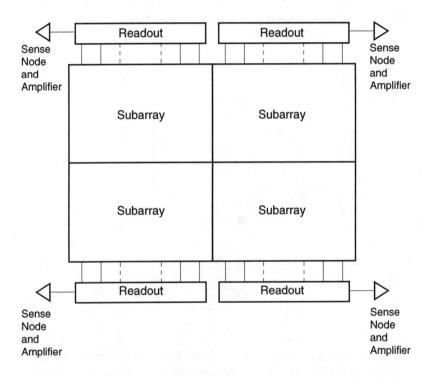

Figure 3-13. A large array divided into four sub-arrays. Each sub-array is read out simultaneously to increase the effective data rate. Very large arrays may have up to 32 parallel outputs. The sub-images are reassembled in electronics external to the device to create the full image.

58 CCD ARRAYS, CAMERAS, and DISPLAYS

3.2.3. FRAME TRANSFER

A frame transfer (FT) imager consists of two almost identical arrays, one devoted to image pixels and one for storage (Figure 3-14). The storage cells are identical in structure to the light sensitive cells but are covered with a metal light shield to prevent any light exposure. After the integration cycle, charge is transferred quickly from the light sensitive pixels to the storage cells. Transfer time to the shielded area depends upon the array size but is typically less than 500 μs. Smear is limited to the time it takes to transfer the image to the storage area. This is much less time than that required by a full frame device.

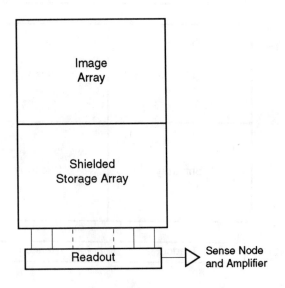

Figure 3-14. Basic architecture for a full frame CCD. Frame transfer devices can also have a split architecture similar to that shown in Figure 3-13.

Since the CCD array is twice the size, it is more complex than a simple full frame device. The array may have a few dummy charge wells between the active area and storage area. These wells collect any anticipated light leakage. Some arrays do not have the integral light shield. These arrays can either be operated in the full frame mode (e.g., 512 x 512) or in the frame transfer mode (e.g., 256 x 512). Here, it becomes the user's responsibility to design the light shield for frame transfer mode.

Example 3-1
SHIELD DESIGN

A camera manufacturer buys a full frame transfer array (512 x 512) but wants to use it in a frame transfer mode (256 x 512). An engineer designs an opaque shield that is placed 1 mm above the array. He is using f/5 optics. What is the usable size of the array if the pixels are 20 μm square and the shield edge is centered on the array?

When placed near the focal plane, the shield edge will create a shadow that is approximately the distance from the focal plane (d = 1 mm) divided by the f-number (F = 5). The full shadow is 200 μm or 10 pixels. One-half of the shadow falls on the active pixels and one-half irradiates the pixels under the shield. This results in a usable array size of 251 x 512. The alignment error (0.5 mm) adds an additional 25 pixels. If the alignment error covers active pixels, the active area is reduced to 226 x 512. If the error covers shielded pixels, the storage area is reduced to 226 x 512. Either way, the usable array is 226 x 512.

Example 3-2
USEFUL ARRAY SIZE

A consumer wants to operate the camera designed in Example 3-1 under relatively low illumination levels. He changes the lens to one whose f-number is 1.25. Will he be satisfied with the imagery?

The lower f-number increases the shadow to 40 pixels. Assuming and alignment error of 0.5 mm, the minimum usable area size has dropped to 211 x 512 with a "soft edge" of 40 pixels. Acceptability of this image size depends upon the application.

3.2.4. INTERLINE TRANSFER

The interline transfer (IL) array consists of photodiodes separated by vertical transfer registers that are covered by an opaque metal shield (Figure 3-15). After integration, the charge generated by the photodiodes is transferred to the vertical CCD registers in about 1 μs and smear is minimized. The main advantage of interline transfer is that the transfer from the active sensors to the shielded storage is quick. There is no need to shutter the incoming light. This is commonly called electronic shuttering. The disadvantage is that it leaves less real estate for the active sensors. The shields act like a Venetian blind that obscures half the information that is available in the scene. The area fill factor may be as low as 20%. Since the detector area is only 20% of the pixel area, the output voltage is only 20% of a detector that would completely fill the pixel area. Microlenses increase the optical fill factor (discussed in Section 3.9, *Microlenses*). Aliasing, with and without microlenses, is discussed in Section 8.3.2., *Detector Array Nyquist Frequency*.

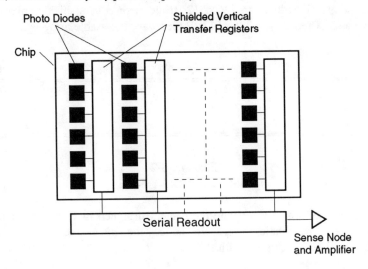

Figure 3-15. Interline transfer architecture. The charge is rapidly transferred to the interline transfer registers. The registers may have three or four gates. Interline transfer devices can also have a split architecture similar to that shown in Figure 3-13 (page 57).

A fraction of the light can leak into the vertical registers. This effect is most pronounced when viewing an ambient scene that has a bright light in it. For professional television applications, the frame interline transfer (FIT) array was developed to achieve even lower smear values (Figure 3-16).

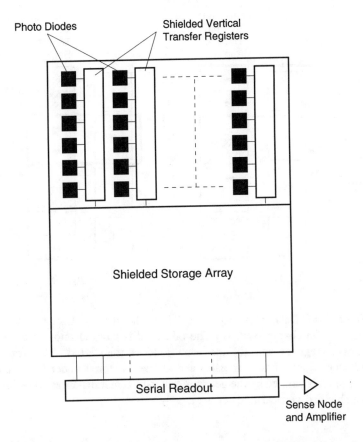

Figure 3-16. Frame-interline transfer architecture. Both the vertical transfer registers and storage area are covered with an opaque mask to prevent light exposure.

A variety of transfer register architectures are available. It can operate with virtual, two, three, or four phases. The number of phases selected depends upon the application. Since interline devices are most often found in consumer camcorder products, most transfer register designs are based upon standard video timing (discussed in Section 5.2.1., *Video Timing*). Figure 3-17 illustrates a four-phase transfer register that stores charge under two gates. With 2:1

interlace, both fields are collected simultaneously but are read out alternately. This is called frame integration. With EIA 170 (formerly called RS 170), each field is read every other 1/60 sec. This allows for a maximum integration time of 1/30 sec for each field.

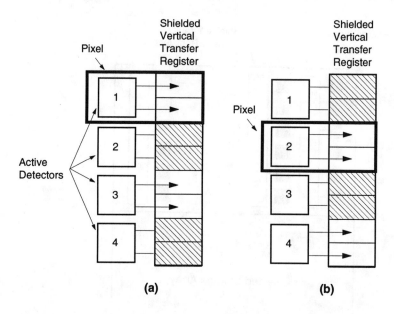

Figure 3-17. Detailed layout of the 2:1 interlaced array. Each photodiode has its own charge well. (a) The odd field is clocked into the vertical transfer register and (b) the even field is transferred. The vertical transfer register has four gates and charge is stored under two wells. The pixel is defined by the detector center-to-center distances. It includes the shielded vertical register area.

Pseudo-interlacing (sometimes called field integration) is shown in Figure 3-18. Changing the gate voltage shifts the image centroid by one-half pixel in the vertical direction. This creates 50% overlap between the two fields. The pixels have twice the vertical extent of standard interline transfer devices and therefore have twice the sensitivity. An array that appears to have 240 elements in the vertical direction is clocked so that it creates 480 lines. However, this reduces the vertical MTF (discussed in Section 8.3.2., *Detector Array Nyquist Frequency*). With some devices, the pseudo-interlace device can also operate in a standard interlace mode (Figure 3-17).

CCD ARRAYS   63

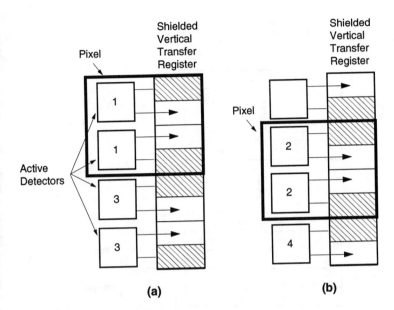

Figure 3-18. Pseudo-interlace. Each photodiode has its own charge well. By collecting charge from alternating active detector sites, the pixel centroid is shifted by one-half pixel. (a) Odd field and (b) even field. The detector numbers are related to the "scan" pattern shown in Figure 3-19.

While any architecture can be used to read the charge, the final video format generally limits the selection. Although the charge is collected in rows and columns, the camera reformats the data into a serial stream consistent with the monitor requirements. This serial stream is related to a fictitious "scan" pattern. Figure 3-19 illustrates the video lines and field-of-view of a standard format imaging system. Figure 3-19a represents the "scan" pattern created by the interlace device shown in Figure 3-17 whereas Figure 3-19b represents the pseudo-interlace "scan" pattern of the device shown in Figure 3-18. Although the dissected scene areas overlap in Figure 3-19b, the video timing is consistent with the video format.

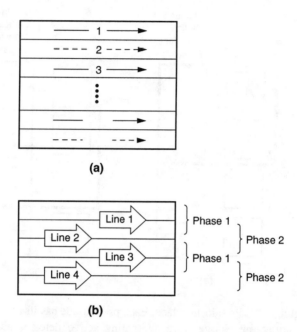

Figure 3-19. Fictitious "scan" patterns. The scan patterns help to visualize how the scene is dissected by the array architecture. (a) Interlace "scanning" achieved with the array shown in Figure 3-17. The solid lines are the odd field and the dashed lines are the even field. (b) Pseudo-interlace "scanning" provided by the array shown in Figure 3-18.

3.2.5. PROGRESSIVE SCAN

Progressive scan simply means noninterlaced or sequential line-by-line "scanning" of the image. CCD arrays do not scan the image but it is convenient to represent the output as if it were scanned. For scientific and industrial applications it is sometimes called slow scan.

The primary advantage of progressive scan is that the entire image is captured at one instant of time. In contrast, interlaced systems collect fields sequentially in time. Any vertical image movement from field-to-field smears the interlaced image in a complex fashion. Horizontal movement serrates vertical lines. With excessive movement, the two fields will have images that are displaced one from the other (Figure 3-20). If a strobe light is used to stop motion, the image will appear only in the field that was active during the light pulse. Because of image motion effects, only one field of data can be used for image processing. This reduces the vertical resolution by 50% and will increase vertical image aliasing.

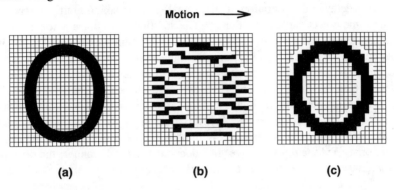

Figure 3-20. Effects of movement. (a) Character moving horizontally, (b) serration that occurs with an interlace system, and (c) output of a progressively scanned array.

Since an entire frame is captured, progressive scan devices do not suffer from same image motion effects seen with interlaced systems. Thus, progressive scan cameras are said to have improved vertical resolution over interlaced systems.

For professional television and consumer video applications, the array aspect ratio, number of elements, and output timing is usually consistent with standard video formats (discussed in Section 5.2.1., *Video Timing*). Although the imagery is collected in a sequential method, camera electronics converts progressively-scanned data into a standard video format. This allows the progressively-scanned camera output to be viewed on any standard display.

For scientific and industrial applications, the progressive output is captured by a frame grabber inside a computer. Once in the computer, the monitor drive electronics automatically reformats the data into the format required by the monitor. The pixel layout and array timing does not need to fit a specified format for these applications.

3.2.6. TIME-DELAY and INTEGRATION

Time-delay and integration (TDI) is analogous to taking multiple exposures of the same object and adding them[10]. The addition takes place automatically in the charge well and the array timing produces the multiple images. Figure 3-21 illustrates a typical TDI application. A simple camera lens inverts the image so the image moves in the opposite direction from the object. As the image is swept across the array, the charge packets are clocked at the same rate. The relative motion between the image and the target can be achieved in many ways. In airborne reconnaissance, the forward motion of the aircraft provides the motion with respect to the ground. With objects on a conveyor belt, the camera is stationary and the objects move. In a document flat bed scanner, the document is stationary but either the camera moves or a scan mirror moves.

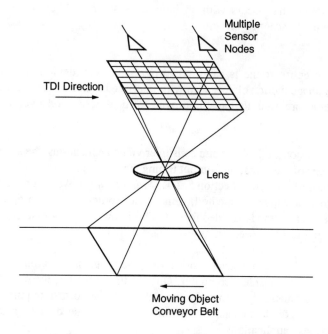

Figure 3-21. Typical TDI operation. The pixel clock rate is matched to the image velocity. This keeps the photoelectrons in phase with the image.

Figure 3-22 illustrates four detectors operating in TDI mode. At time T_1 the image is on the first detector and creates a charge packet. At time T_2, the image has moved to the second detector. Simultaneously, the pixel clock moved the charge packet to the well under the second detector. Here, the image creates additional charge that is added to the charge created by the first detector. The charge (signal) increases linearly with the number of detectors in TDI. Noise also increases, but as the square root of the number of TDI elements, N_{TDI}. This results in an SNR improvement of $\sqrt{N_{TDI}}$. The well capacity limits the maximum number of TDI elements that can be used.

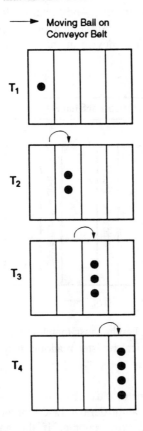

Figure 3-22. TDI concept. At time T_1, the image is focused onto the first detector element. At T_2 the image has moved to the next detector. Simultaneously, the charge packet is clocked down to the next pixel site. The image creates additional charge that is added to the stored packet and the signal increases linearly with the number of TDI elements. After four stages, the signal increases fourfold.

68 CCD ARRAYS, CAMERAS, and DISPLAYS

For this concept to work, the charge packet must always be in synch with the moving image. If there is a mismatch between the image scan velocity and pixel clock rate, the output is smeared and this adversely affects the in-scan MTF (discussed in Section 10.3.3., *TDI*). The accuracy to which the image velocity is known limits the number of useful TDI stages.

As shown in Figure 3-21, to image the entire target, the array height must be greater than the image height. Although Figure 3-21 illustrates two readout registers, an array may have multiple readout registers. Array sizes may be as large as 2048 by 32 TDI elements. Additional readouts are required to avoid charge well saturation. The multiple readout outputs may be summed in electronics external to the TDI device or may be placed in a horizontal serial register for summing (Figure 3-23).

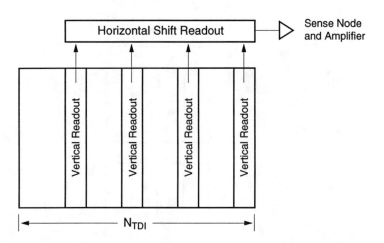

Figure 3-23. The array height must match the image height. To avoid saturation, multiple vertical register readouts may be used.

The vertical readout register must clock out the data before the next line is loaded into the register. The amplifier bandwidth or analog-to-digital conversion time limits the pixel rate. For example, if the amplifier can support 10^7 pixels/sec and the register serves 500 vertical pixels, then the maximum (in-scan) line rate is 20,000 lines/sec. If the amplifier can only handle 2×10^6 pixels/sec and the line rate is maintained at 20,000 lines/sec, then the vertical transfer register must be divided into five registers - each connected to 100 pixels. These registers and associated outputs operate in parallel (Figure 3-24).

CCD ARRAYS 69

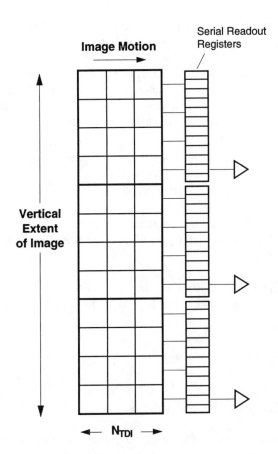

Figure 3-24. The vertical direction may be divided into sub-arrays to increase the line rate. The transfer register typically is a three-phase device.

TDI improves photoresponse nonuniformity (discussed in Section 4.2.6., *Pattern Noise*). Variations in individual detector responsivities in the TDI direction are averaged by $1/N_{TDI}$. If the array is serviced by multiple sense nodes (and therefore different on-chip and off-chip amplifiers), the displayed image of the sub-arrays may vary in contrast and level. This is due to differences in amplifier gains and level adjustments.

70 CCD ARRAYS, CAMERAS, and DISPLAYS

Multiple outputs are used to increase the effective pixel rate. Each output is amplified by a separate amplifier that has unique gain characteristics. If the gains are poorly matched, the image serviced by one amplifier will have a higher contrast than the remaining image. In TDI devices, streaking occurs when the average responsivity changes from column-to-column[11]. Careful selection of arrays will minimize these nonuniformities. They can also be minimized though processing techniques employed external to the device.

Example 3-3
TDI LINE RATE

The pixels in a TDI are 15 μm square. The object is moving at 2 m/sec and the object detail of interest is 150 μm. What is the lens magnification and what is the pixel clock rate?

Each pixel must view 150 μm on the target. Since the pixel size is 15 μm, the lens magnification must be 150/15 = 10. The TDI pixel clock rate is (2 m/sec)/150 μm = 133 Khz.

3.2.7. COLOR FILTER ARRAYS

The subjective sensation of color can be created from three primary colors. By adjusting the intensity of each primary (additive mixing), a full gamut (rainbow) of colors is experienced. This approach is used on all color displays. They have red, green, and blue phosphors that, when appropriately excited, produce a wide spectrum of perceived colors. The CIE committee standardized a color perception model for the human observer in 1931. The literature is rich with visual data that forms the basis for color camera design[12].

The "color" signals sent to the display must be generated by three detectors, each sensitive to a primary or its complement. For high-quality color imagery, three separate detectors are used whereas for consumer applications, a single array is used. The detectors are covered with different filters that, with the detector response, can approximate the primaries or the complements (Figure 3-25). These arrays are called colored filtered arrays (CFAs).

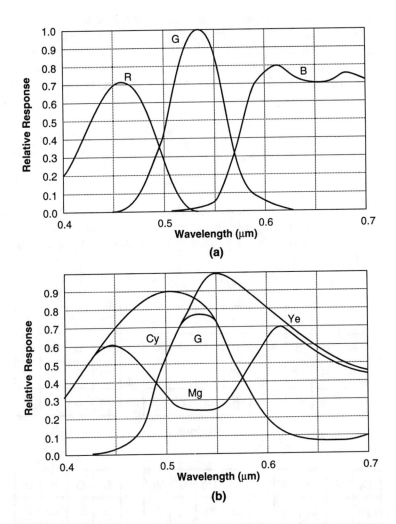

Figure 3-25. Desired spectral response for the three primaries and their complements. (a) The primaries are red, green, and blue (R, G, B) and (b) their complements are yellow, cyan, and magenta (Ye, Cy, Mg).

While CFAs can have filters with the appropriate spectral transmittance to create R, G, and B, it is more efficient to create the complementary colors. Complementary filters have higher transmittances than the primary filters. The luminance channel is identical to the green channel and is label Y. $W = R + G + B$ and represents "white." It is transparent (i.e., no color filter).

72 CCD ARRAYS, CAMERAS, and DISPLAYS

The primary additive colors are red, green, and blue (R, G, B) and their complimentary colors are yellow, cyan and magenta (Ye, Cy, Mg). A linear set of equations relates the primaries to their complements. These equations are employed (called matrixing) in cameras to provide either or both output formats (discussed in Section 5.3.2. *Color Correction*). Matrixing can convert RGB into other color coordinates:

$$Ye = R + G = W - B \qquad (3\text{-}4a)$$

$$Mg = R + B = W - G \qquad (3\text{-}4b)$$

$$Cy = G + B = W - R \qquad (3\text{-}4c)$$

The arrangement of the color filters for a single array system is either a stripe or mosaic pattern (Figure 3-26). The precise layout of the mosaic varies by manufacturer. One basic CFA patent[13] was granted to Bryce E. Bayer at Eastman Kodak in 1976.

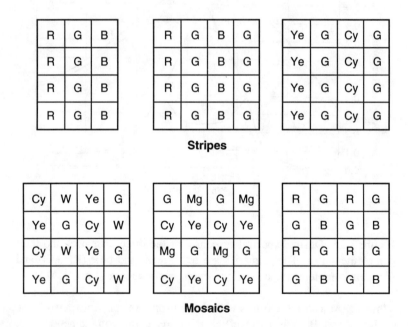

Figure 3-26. Representative stripe and mosaic arrays. Although shown as a full frame for clarity, they typically are interline transfer devices. The layout depends upon the manufacturer's philosophy and cleverness in reducing color aliasing.

In many sensors, the number of detectors devoted to each color is different. The basic reason is that the human visual system (HVS) derives its detail information primarily from the green portion of the spectrum. That is, luminance differences are associated with green whereas color perception is associated with red and blue. The HVS requires only moderate amounts of red and blue to perceive color. Thus many sensors have twice as many green as either red or blue detector elements. An array that has 768 horizontal elements may devote 384 to green, 192 to red, and 192 to blue. This results in an unequal sampling of the colors. A birefringent crystal (discussed in Section 10.3.4., *Optical Anti-alias Filter*), inserted between the lens and the array, accommodates the different sampling rates (discussed in Section 8.3.2., *Detector Array Nyquist Frequency*). The video signal from the CFA is embedded in the architecture of the CFA. The color video from a single chip CFA must be decoded (unscrambled) to produce usable R, G, B signals.

3.3. DARK CURRENT

The CCD output is proportional to the exposure, $L_q(\lambda,T)t_{INT}$. The output can be increased by increasing the integration time and long integration times are generally used for low-light-level operation. However, this approach is ultimately limited by dark current leakage that is integrated along with the photocurrent. With a large pixel (24 μm square), a dark current density of 1000 pa/cm^2 produces 36,000 electrons/pixel/sec. If the device has a well capacity of 360,000 electrons, the well fills in 10 seconds. The dark current is only appreciable when t_{INT} is long. It is not usually a problem for consumer applications but is a concern for scientific applications.

A critical design parameter is the dark noise reduction. There are three main sources of dark current: (1) thermal generation and diffusion in the neutral bulk material, (2) thermal generation in the depletion region, and (3) thermal generation due to surface states. Dark current densities vary significantly among manufacturers with values ranging from 0.1 na/cm^2 to 10 na/cm^2 in silicon CCDs. Dark current due to thermally generated electrons can be reduced by cooling the device. Surface state dark current is minimized with multi-phase pinning[14,15] (MPP).

The number of thermally generated dark current electrons is

$$n_{DARK} = \frac{J_D A_D t_{INT}}{q} \quad (3\text{-}5)$$

J_D is the dark current density and it is dependent upon the temperature:

$$J_D \approx k e^{-\frac{E_G}{\alpha_G kT}} \quad (3\text{-}6)$$

where E_G is the band gap and $1 < \alpha_G < 2$. With TDI, the pixels in a TDI column are summed and the dark current increases:

$$n_{DARK} = \frac{J_D A_D t_{INT} N_{TDI}}{q} \quad (3\text{-}7)$$

In principle, dark current density can be made negligible with sufficient cooling. Dark current density decreases approximately twofold for every 7 to 8° C drop in array temperature. Conversely, the dark current density increases by a factor of two for every 7 to 8° C increase in array temperature. Cooling probably is only worthwhile for scientific applications.

The most common cooler is a thermoelectric cooler (TEC or TE cooler). Thermoelectric coolers are Peltier devices driven by an electric current that pumps heat from the CCD to a heat sink. The heat sink is cooled by passive air, forced air, or forced liquid (usually water). The final array temperature depends upon the amount of heat generated in the array, the cooling capacity of the thermoelectric device, and the temperature of the TEC heat sink. Peltier devices can cool to about -60°C. Thermoelectric coolers can be integrated into the same package with the array.

Although liquid nitrogen (LN, LN2, or LN$_2$) is at -200° C, the typical operating temperature is -120° C to -60° C because the CTE and quantum efficiency drop at very low temperatures. Condensation is a problem and the arrays must be placed in an evacuated chamber or a chamber that is filled with a dry atmosphere. Dark current can approach[16] 3.5 electrons/pixel/sec at -60°C and 0.02 electrons/pixel/hour at -120°C.

Multi-pinned phasing technology CCDs can operate at room temperature with reduced dark current[14,15]. MPP devices have dark current density levels as low as 2 pa/cm^2 with a high value of 1 na/cm^2. Typically, MPP devices have charge wells that are 2-3 times smaller than conventional devices and these devices are more likely to saturate. Some devices operate in either mode. MPP is the most common architecture used in scientific and medical applications.

The output amplifier constantly dissipates power. This results in local heating of the silicon chip. Since dark current is dependent upon the temperature, a small variation in the device temperature profile produces a corresponding profile in dark current. To minimize this effect, the output amplifier is often separated by several isolation pixels to remotely locate the amplifier with respect to the active sensors.

☞

Example 3-4
DARK CURRENT

A CCD array has a well size of 150,000 electrons, dark current density of 10 pa/cm^2, and detector size 8 μm square. The advertised dynamic range is 80 dB. Is the dark current noise significant? Assume that the integration time is 16.67 msec.

Using Equation 3-5, the number of dark electrons is

$$n_{DARK} = \frac{\left(10 \times 10^{-9} \frac{amps}{cm^2}\right)(0.0167 \, sec)(8 \times 10^{-4} cm)^2}{1.6 \times 10^{-19} \frac{coul}{electron}} = 668 \ e^- \quad (3\text{-}8)$$

Assuming Poisson statistics, the dark current shot noise is $\sqrt{668}$ or 26 electrons rms. The dynamic range is the charge well capacity divided by the total noise level in the dark. With a dynamic range of 10000:1, the noise floor is 15 electrons. Since the noise variances add, the total noise is:

$$\langle n_{SYS} \rangle = \sqrt{26^2 + 15^2} = 30 \ e^- \ rms \quad (3\text{-}9)$$

Therefore, the dark current will affect the measured SNR at low levels.

Example 3-5
ARRAY COOLING

The dark current density decreases by a factor of two for every 8 °C drop in temperature. What should the array temperature be so that the dark current noise is 5 electrons rms for the array described in example 3-4?

The dark current noise must be reduced by a factor of $5/26 = 0.192$ and the dark current must be reduced by $(0.192)^2$ or 0.0369. The doubling factor provides

$$\frac{n_{DARK-COOL}}{d_{DARK-AMBIENT}} = 2^{\left(-\frac{T_{AMBIENT} - T_{COOL}}{8}\right)} \qquad (3\text{-}10)$$

Then, $T_{AMBIENT} - T_{COOL} = 38.1°\text{C}$. That is, the array temperature must be 38.1° C below the temperature at which the dark current density was originally measured. Although this varies by manufacturer, many use $T_{AMBIENT} = 20°\text{C}$.

3.4. DARK PIXELS

Many arrays have "extra" photosites at the end of the array. These photosites are shielded and used to establish a reference dark current level (or dark signal). The average value of the dark current pixels is subtracted from the active pixels leaving only photo-generated signal. For example, if the output of the dark pixels is 5 mV, then 5 mV is subtracted from the signal level of each active pixel.

The number of dark elements varies with device and manufacturer and ranges from a few to 25. The total number of elements specified by the device manufacturer (e.g., 755 x 484) usually includes the dark pixels.

Figure 3-27a illustrates the output from a single charge well. Light leakage may occur at the edge of the shield (see Example 3-1) and partially fill a few wells. Because of this light leakage, a few pixels (called dummy or isolation pixels) are added to the array. Figure 3-27b illustrates the output of three dark reference detectors, three isolation detectors, and three active pixels. The average value of the three dark pixels is subtracted from every active pixel value.

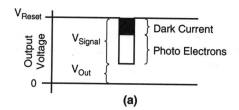

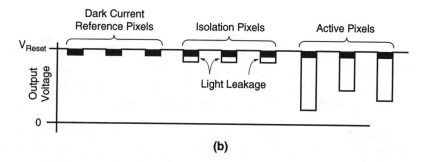

Figure 3-27. Typical output. (a) Single pixel and (b) active line with isolation and dark pixels. The varying voltage on the active pixels is proportional to the scene intensity. The dark pixels are covered with an opaque mask. Light can leak around the mask and partially fill the isolation pixel wells.

While this process is satisfactory for consumer camcorder and machine vision use, it may be unacceptable for scientific applications. Dark pixels have slightly different dark current than the active pixels. Furthermore, the dark current value varies from pixel to pixel. This variation is fixed pattern noise (discussed in Section 4.2.6., *Pattern Noise*). Removing the average value does not remove the variability.

In critical scientific applications, only the dark value from a pixel may be removed from that pixel. That is, the entire array is covered (shutter in place) and the individual pixel values are stored in a matrix. Then the shutter is removed and the stored values are subtracted, pixel by pixel, as a matrix subtraction. This method insures that the precise dark voltage is subtracted from each pixel. This, of course, increases computational complexity. Removal of dark current values allows maximum use the analog-to-digital converter dynamic range. It cannot remove the dark current shot noise that is always present. Dark current shot noise is reduced by cooling.

Additional isolation pixels may exist between the readout register and the sense node amplifier. These pixels reduce any possible interaction between the amplifier temperature and first column pixel dark current. They also reduce amplifier noise interactions.

3.5. ANTI-BLOOMING DRAIN

When a potential well fills, charge spills over into adjacent pixels on the same column resulting in an undesirable overload effect called blooming. Channel stops prevent spill over onto adjacent columns. Overflow drains or anti-blooming drains prevent spill over. The drain can be attached to every pixel or may only operate on a column of pixels. With interline transfer devices, the structure is also called a vertical overflow drain (VOD).

The overflow drain consumes pixel real estate and thus decreases the light sensitive area. Light incident onto the anti-blooming structure generates charge that is instantly removed. If the drain is buried, the long wavelength quantum efficiency is reduced since the long wavelength charge is generated below the wells and the buried drain.

In a perfect system, the output is linearly related to the input up to the anti-blooming drain limit. In real arrays, because of imperfect drain operation, a knee is created[17] (Figure 3-28). The response above the knee is called knee slope or knee gain. While scientific cameras operate in a linear fashion, there is no requirement to do so with consumer cameras. The knee is a selling feature in the consumer marketplace. The advantage of the knee is that it provides some response above the white clipping level. Each CCD manufacturer claims various competitive methods of handling wide scene contrasts.

The anti-blooming drain can also be used for exposure control by holding the overflow drain bias to a maximum level and clocking the adjacent gate. The exposure control can work at any speed. The drain, acting as an electronic shutter, may be used synchronously or asynchronously. Asynchronous shuttering is important for use in processes where the motion of objects does not match the array frame timing or in high-light-level applications.

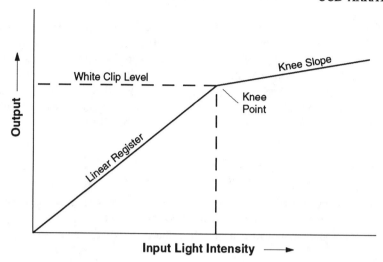

Figure 3-28. Characteristic knee created by the vertical overflow drain.

3.6. CHARGE TRANSFER EFFICIENCY

As the charge packet moves from storage site to storage site, a few electrons are left behind. The ability to transfer all the charge is given by the charge transfer efficiency (CTE). As the number of transfers increases, the charge transfer efficiency must also increase to maintain signal strength.

The number of transfers across the full array is the number of pixels multiplied by the number of gates. For example, for an array that is 1000 x 1000, the charge packet farthest from the sense node must travel 2000 pixels. If four gates (four-phase device) are used to minimize cross talk, the farthest charge packet must pass through 8000 wells (neglecting isolation pixels). The total number increases as the number of isolation pixels increases. The net efficiency varies with the target location. If the target is in the center of the array, then it only experiences one-half of the maximum number of transfers on the array. If the target is at the leading edge, it is read out immediately with virtually no loss of information. The loss of information is described by the CTE MTF (discussed in Section 10.3.2., *Charge Transfer Efficiency*).

If many transfers exist (e.g., 8,000) the transfer efficiency must be high (e.g., 0.999995) so that the net efficiency is reasonable ($0.999995^{8000} = 0.96$). Arrays for consumer applications typically have charge transfer efficiencies greater than 0.9999 and CTEs approach 0.999999 for scientific grade devices.

80 CCD ARRAYS, CAMERAS, and DISPLAYS

A fraction of the charge, as specified by the charge *in*efficiency, is left behind. After the first transfer, ϵ (ϵ is the CTE) remain in the leading well and (1-ϵ) enters the first trailing well. After the second transfer, the leading well transfers ϵ of the previous transfer and loses (1-ϵ) to the trailing well. The trailing well gains the loss from the first well and adds it to the charge transferred from the trailing well in the last go-around. Charge is not lost, just rearranged. The fractional values follow a binomial probability distribution given by

$$P_R = \binom{N}{R} \epsilon^R (1 - \epsilon)^{N-R} \qquad (3\text{-}11)$$

where N is the number of transfers. For the first well, R = N. For the second well, R = N - 1 and so on. Let x be the first well and y be the first trailing well. The fractional amount of charge left in the leading well after N transfers is

$$x_N = \epsilon^N \qquad (3\text{-}12)$$

The fractional amount of charge appearing in the trailing well after N transfers is

$$y_N = N \epsilon^{N-1} (1 - \epsilon) \qquad (3\text{-}13)$$

The fractional signal strength, x_N, is plotted in Figure 3-29. The first trailing well fractional strength, y_N, is plotted in Figure 3-30. Most of the electrons lost from the first well appear in the second and third wells. For high CTE arrays, a negligible amount appears in successive wells.

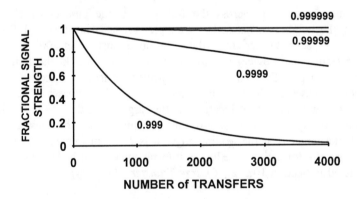

Figure 3-29. Fractional output as a function of transfer efficiency and number of transfers. Large arrays require high CTE.

CCD ARRAYS 81

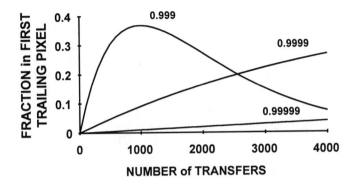

Figure 3-30. Fractional output for the first trailing charge well. With low CTEs, as the number of transfers increases, more charge spills into the second trailing well. Simultaneously, the charge in the first trailing well decreases.

CTE is not a constant value but depends upon the charge packet size. For very low charge packets, the CTE decreases due to surface state interactions. This is particularly bothersome when the signal-to-noise ratio is low and the array is very large (typical of astronomical applications). The CTE also decreases near saturation due to charge spill effects.

CTE effects in large linear arrays can be minimized by using alternate readouts. In Figure 3-31, the alternate readouts reduce the number of transfers by a factor of two. Adjacent pixels suffer the same number of transfers so that the signal is preserved at any particular location.

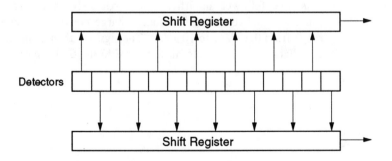

Figure 3-31. For long arrays, the MTF can be increased by alternating the readouts or by reading out sub-arrays (see Figure 3-13, page 57).

There are three methods to measure CTE: current injection, optical injection, and x-ray injection. Since most devices do not have current injection terminals, optical and x-ray injection are used more often. With the optical technique, either a row of detectors or a single detector is illuminated. The ratio of the output of the first illuminated pixel to the average value of trailing pixels is measured. The CTE is found using Equation 3-11. The optical technique requires a spot[18] whose diameter is less than a pixel width. This can only be achieved with carefully designed optics that is critically aligned. On the other hand, x-rays[19] generate a known number of electrons in a localized area that is much smaller than a pixel. The x-ray technique probably provides the highest accuracy.

3.7. CHARGE CONVERSION (OUTPUT STRUCTURE)

Charge is converted to a voltage by a floating diode or floating diffusion. Figure 3-32 illustrates the last two gates of a four-phase system. The low voltage gate allows charge transfer only when gate 4 goes to zero. The diode, acting as a capacitor, is precharged at a reference level. The capacitance, or sense node, is partially discharged by the amount transferred. The difference in voltage between the final status of the diode and its precharged value is linearly proportional to the number of electrons, n_e. The signal voltage after the source follower is

$$V_{SIGNAL} = V_{RESET} - V_{OUT} = n_e \frac{Gq}{C} \qquad (3\text{-}14)$$

The gain, G, of a source follower amplifier is approximately one. q is the electronic charge and is equal to 1.6×10^{-19} coul. Charge conversion values (Gq/C) typically range from $0.1\ \mu V/e^-$ to $10\ \mu V/e^-$. The signal is then amplified, processed, and digitally encoded by electronics external to the CCD sensor.

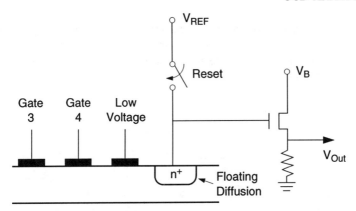

Figure 3-32. Output structure. The charge on the floating diffusion is converted to a voltage by the diffusion capacitance. The low voltage gate, which operates in a manner similar to a virtual well, allows charge transfer when the voltage on gate 4 is zero (see Figure 3-10, page 52).

3.8. CORRELATED DOUBLE SAMPLING

After a charge packet is read, the floating diffusion capacitor is reset before the next charge packet arrives. The uncertainty in the amount of charge remaining on the capacitor following reset appears as a voltage fluctuation (discussed in Section 4.2.2., *Reset Noise*).

Correlated double sampling (CDS) assumes that the same voltage fluctuation is present on both the video and reset levels. That is, the reset and video signals are correlated in that both have the same fluctuation. The CDS circuitry may be integrated into the array package and this makes processing easier for the end user. For this reason it is included in this chapter.

CDS may be performed[20] in the analog domain by a clamping circuit, with a delay-line processor, or by the circuit illustrated in Figure 3-33. When done in the digital domain, a high-speed analog-to-digital converter (ADC) is required that operates much faster than the pixel clock. Limitations of the ADC may restrict the use of CDS in some applications. The digitized video and reset pulses are subtracted and then clocked out at the pixel clock rate.

The operation of the circuit is understood by examining the timing signals (Figure 3-34). When the valid reset pulse is high, the reset switch closes and the signal (reset voltage) is stored on capacitor C_R. When the valid video pulse is high, the valid video switch closes and the signal (video voltage) is stored on capacitor C_V. The reset voltage is subtracted from the video voltage in a summing amplifier to leave the CDS corrected signal. CDS can also reduce amplifier 1/f noise.

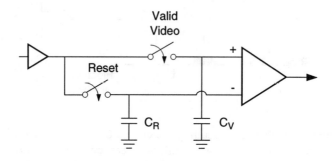

Figure 3-33. Correlated double sampling circuitry. The switches close according to the timing shown in Figure 3-34.

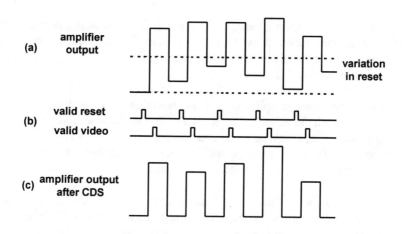

Figure 3-34. Correlated double sampling. (a) Amplifier output signal before CDS, (b) timing to measure video and reset levels, and (c) the difference between the video and reset levels after CDS. The dashed lines indicate the variation in the reset level.

3.9. MICROLENSES

Optical fill factor may be less than 100% due to manufacturing constraints in full transfer devices. In interline devices, the shielded storage area reduces the fill factor to less than 20%. Microlens assemblies (also called microlenticular arrays or lenslet arrays) increase the effective optical fill factor (Figure 3-35). But it may not reach 100% due to slight misalignment of the microlens assembly, imperfections in the microlens itself, nonsymmetric shielded areas, and transmission losses. As shown in the camera formula (Equation 2-15, page 24), the output is directly proportional to the detector area. Increasing the optical fill factor with a microlens assembly increases the effective detector size and, therefore, the output voltage.

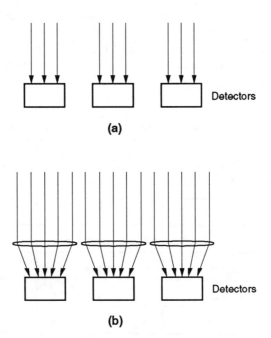

Figure 3-35. Optical effect of a microlens assembly. (a) With no microlens, a significant amount of photon flux is not detected. (b) The microlens assembly can image nearly all the flux onto the detector when a high camera f-number is used. The effective of the microlens decreases as the f-number decreases. These lenslets can either be grown on the array during the fabrication process or manufactured out of a material such as quartz and placed on the array surface during packaging.

86 CCD ARRAYS, CAMERAS, and DISPLAYS

The photosensitive area is below the gate structure and the ability to collect the light depends upon gate thickness. The cone of light reaching the microlens depends upon the f-number of the primary camera lens. Figure 3-35 illustrates nearly parallel rays falling on the microlens. This case is encountered with high f-number camera lens systems. Low f-number primary camera lenses increase the cone angle and the effective fill factor decreases with decreasing f-number[21]. Microlenses are optimized for most practical f-numbers. As the array size grows, off axis detectors do not obtain the same benefit as on-axis detectors[22].

3.10. VIDEO CHIP SIZE

Historically, vidicon vacuum tubes used for professional television applications were specified by the tube diameter. To minimize distortion and nonuniformities within the tube, the recommended[23] image size was considerably less than the overall tube diameter. When CCDs replaced the tubes, the CCD industry maintained the image sizes but continued to use the tube nomenclature (Table 3-1).

Table 3-1
CCD ARRAY SIZE for STANDARD FORMATS

NOMINAL CAMERA SIZE	STANDARDIZED ARRAY SIZE (H X V)	CCD ARRAY DIAGONAL
1 inch	12.8 mm x 9.6 mm	16 mm
2/3 inch	8.8 mm x 6.6 mm	11 mm
1/2 inch	6.4 mm x 4.8 mm	8 mm
1/3 inch	4.8 mm x 3.6 mm	6 mm
1/4 inch	3.2 mm x 2.4 mm	4 mm

Although each manufacturer supplies a slightly different array size and pixel size, nominal sizes for a 640 x 480 array are given in Table 3-2. These pixels tend to be square. However, with interline transfer devices, approximately one-half of the pixel width is devoted to the shielded vertical transfer register. That is, the detector active width is one-half of the pixel width. Thus the active area of the pixel is rectangular in interline transfer devices. This asymmetry does not appear to significantly affect image quality in consumer video products.

Table 3-2
NOMINAL PIXEL SIZE for a 640 x 480 ARRAY
Sizes vary by manufacturer. Detector sizes are smaller.

NOMINAL CAMERA SIZE	NOMINAL PIXEL SIZE (H X V)
1 inch	20 μm x 20 μm
2/3 inch	13.8 μm x 13.8 μm
1/2 inch	10 μm x 10 μm
1/3 inch	7.5 μm x 7.5 μm
1/4 inch	5 μm x 5 μm

The decrease in optical format is related to cost. The price of CCD arrays is mainly determined by the cost of processing semiconductor wafers. As the chip size decreases, more devices can be put on a single wafer and this lowers the price of each individual device. The trend of going to smaller devices will probably continue as long as the optical and electrical performance of the imagers do not change. However, smaller pixels reduce the charge well size. For a fixed flux level and lens f-number, the smaller arrays have reduced sensitivity.

Smaller chips make for smaller cameras. However, to maintain resolution, pixels can only be made so small. Here, the tradeoff is among pixel size, optical focal length, and overall chip size.

3.11. DEFECTS

Large arrays sometimes contain defects. These are characterized in Table 3-3. Arrays with a few defects are more expensive than arrays with a large number. Depending upon the application, the location of the defects may be important. For example, the center ("sweet spot") must be fully operational with increasing defects allowed as the periphery is approached.

Table 3-3
ARRAY DEFECTS

DEFECTS	DEFINITION
Point defect	A pixel whose output deviates by more than 6% compared to adjacent pixels when illuminated to 70% saturation.
Hot point defects	Pixels with extremely high output voltages. Typically a pixel whose dark current is 10 times higher than the average dark current.
Dead pixels	Pixels with low output voltage and/or poor responsivity. Typically a pixel whose output is one-half of the others when the background nearly fills the wells.
Pixel traps	A trap interferes with the charge transfer process and results in either a partial or whole bad column (either all white or all dark).
Column defect	Many (typically 10 or more) point defects in a single column. May be caused by pixel traps.
Cluster defect	A cluster (grouping) of pixels with point defects.

3.12. REFERENCES

1. J. Janesick, T. Elliott, R. Winzenread, J. Pinter, and R. Dyck, "Sandbox CCDs," in *Charge-Coupled Devices and Solid State Optical Sensors V*, M. M. Blouke, ed., SPIE Proceedings Vol. 2415, pp. 2-42 (1995).
2. J. Janesick and T. Elliott, "History and Advancement of Large Area Array Scientific CCD Imagers," in *Astronomical Society of the Pacific Conference Series, Vol. 23, Astronomical CCD Observing and Reduction*, pp. 1-39, BookCrafters (1992).
3. M. J. Howes and D. V. Morgan, eds., *Charge-Coupled Devices and Systems*, John Wiley and Sons, New York, NY (1979).
4. C. H. Sequin and M. F. Tompsett, *Charge Transfer Devices*, Academic Press, New York, NY (1975).
5. E. S. Yang, *Microelectronic Devices*, McGraw-Hill, NY (1988).
6. E. L. Dereniak and D. G. Crowe, *Optical Radiation Detectors*, John Wiley and Sons, New York, NY (1984).
7. A. J. P. Theuwissen, *Solid-State Imaging with Charge-Coupled Devices*, Kluwer Academic Publishers, Dordrecht, The Netherlands (1995).
8. A. J. P. Theuwissen, *Solid-State Imaging with Charge-Coupled Devices*, pp. 317-348, Kluwer Academic Publishers, Dordrecht, The Netherlands (1995).

CCD ARRAYS 89

9. S. G. Chamberlain, S. R. Kamasz, F. Ma, W. D. Washkurak, M. Farrier, and P.T. Jenkins, "A 26.3 Million Pixel CCD Image Sensor," in *IEEE Proceedings of the International Conference on Electron Devices*, pp. 151-155, Washington, D.C. December 10, 1995.
10. H.-S. Wong, Y. L. Yao, and E. S. Schlig, "TDI Charge-Coupled Devices: Design and Applications," *IBM Journal Research Development*, Vol. 36(1), pp. 83-106 (1992).
11. T. S. Lomheim and L. S. Kalman, "Analytical Modeling and Digital Simulation of Scanning Charge-Coupled Device Imaging Systems," in *Electro-Optical Displays*, M. A. Karim, ed., pp. 551-560, Marcel Dekker, New York (1992).
12. See, for example, J. Peddie, *High-Resolution Graphics Display Systems*, Chapter 2, Windcrest/McGraw-Hill, New York, NY (1993) or A. R. Robertson and J. F. Fisher "Color Vision, Representation, and Reproduction," in *Television Engineering Handbook*, K. B. Benson, ed. Chapter 2, McGraw Hill, New York, NY (1985).
13. B. E. Bayer, US patent #3,971,065 (1976).
14. J. Janesick, "Open Pinned-Phase CCD Technology," in *EUV, X-ray, and Gamma Ray Instruments for Astronomy and Atomic Physics*, C. J. Hailey and O. H. Siegmund, eds., SPIE Proceedings Vol. 1159, pp 363-373 (1989).
15. J. Janesick, T. Elliot, G. Fraschetti, S. Collins, M. Blouke, and B. Corey, "CCD Pinning Technologies," in *Optical Sensors and Electronic Photography*, M. M. Blouke, ed., SPIE Proceedings Vol. 1071, pp. 153-169 (1989).
16. J. R. Janesick, T. Elliott, S. Collins, M. M. Blouke, and J. Freeman, "Scientific Charge-coupled Devices," *Optical Engineering*, Vol. 26(8), pp. 692-714 (1987).
17. S. Kawai, M. Morimoto, N. Mutoh, and N. Teranishi, "Photo response Analysis in CCD Image Sensors with a VOD Structure," *IEEE Transactions on Electron Devices*, Vol. 40(4), pp. 652-655 (1995).
18. G. Wan, X. Gong, and Z. Luo, "Studies on the Measurement of Charge Transfer Efficiency and Photoresponse Nonuniformity of Linear Charge-coupled Devices," *Optical Engineering*, Vol. 34 (11), pp.3254-3260 (1995).
19. J. Janesick, T. Elliott, R. Winzenread, J. Pinter, and R. Dyck, "Sandbox CCDs," in *Charge-coupled Devices and Solid State Optical Sensors V*, M. M. Bloke, ed., SPIE Proceedings Vol. 2415, pp. 2-42, (1995).
20. A. J. P. Theuwissen, *Solid-State Imaging with Charge-Coupled Devices*, pp. 228-231, Kluwer Academic Publishers, Dordrecht, The Netherlands (1995).
21. J. Furukawa, I. Hiroto, Y. Takamura, T. Walda, Y. Keigo, A. Izumi, K. Nishibori, T. Tatebe, S. Kitayama, M. Shimura, and H. Matsui, "A ⅓-inch 380k Pixel (Effective) IT-CCD Image Sensor," *IEEE Transactions on Consumer Electronics*, Vol. CE-38(3), pp. 595-600 (1992).
22. M. Deguchi, T. Maruyama, F. Yamasaki, T. Hamamoto, and A. Izumi, "Microlens Design Using Simulation Program for CCD Image Sensor," *IEEE Transactions on Consumer Electronics*, Vol. CE-38(3), pp. 583-589 (1992).
23. R. G. Neuhauser, "Photosensitive Camera Tubes and Devices," in *Television Engineering Handbook*, K. B. Benson, ed., pg. 11.34, McGraw-Hill, New York, NY (1986).

4
ARRAY PERFORMANCE

The most common array performance measures are spectral response, minimum signal, maximum signal, dynamic range, pixel-to-pixel uniformity, and output conversion gain. Noise determines the minimum detectable signal. Since the array is the basic building block of the camera, camera terminology is often used for array specifications. Additional performance measures exist at the system level. These include the noise equivalent irradiance (NEI) and noise equivalent differential reflectance (NE$\Delta\rho$) and are discussed in Chapter 5, *Camera Performance*.

Full characterization includes quantifying the noise floor, charge transfer efficiency, spectral quantum efficiency, full well capacity, linearity, pixel nonuniformity, signal-to-noise ratio, and dynamic range. While all these metrics may be necessary for the most critical scientific applications, consumers are interested in read noise, charge well capacity, and responsivity.

The signal is modified by the spatial response of the optics and detector whereas the noise originates in the detector and electronics. Noise characteristics may be different from signal characteristics. System design requires optimizing the signal-to-noise ratio. Sensitivity should not the only performance parameter for selecting a system.

The magnitude of each noise component must be quantified and its effect on system performance must be understood. Noise sources may be a function of the detector temperature. Predicted system performance may deviate significantly from actual performance if significant 1/f noise or other noise is present. There is a myriad of factors involved in system optimization. It is essential to understand what limits the system performance so that intelligent improvements are made.

Janesick[1] uses three methods to characterize CCD performance: (1) photon transfer, (2) x-ray transfer, and (3) photon standard. The photon transfer technique (described in this chapter) appears to be the most valuable in calibrating, characterizing and optimizing arrays for consumer and industrial applications. Additional information, necessary for characterizing scientific arrays, is obtained from the x-ray transfer and photon standard methods.

Soft x-ray photons produce a known number of electron-hole pairs in a localized area of the CCD. Since the number and location of electron hole-pairs are known precisely, it is easy to determine the charge transfer efficiency. With the photon standard technique, the CCD array views a source with a known spectral distribution. Using calibrated spectral filters or a monochrometer, the spectral quantum efficiency is determined.

The symbols used in this book are summarized in the *Symbol List* (page xiv) and it appears after the *Table of Contents*.

4.1. SIGNAL

Device specifications depend, in part, upon the application. Arrays for consumer video applications may provide the responsivity in units of volts/lux. For scientific applications, the units may be in electrons/(Joules-cm^{-2}) or, if a digital output is available, DN/(Joules-cm^{-2}) where DN refers to a digital number. For example, in an 8-bit system the digital numbers range from zero to 255. These units are incomplete descriptors unless the device spectral response and source spectral characteristics are furnished.

The maximum output occurs when the charge wells are filled. The exposure that produces this value is the saturation equivalent exposure (SEE). With this definition, it is assumed that the dark current produces an insignificant number of electrons so that the only photoelectrons fill the well.

Sensitivity suggests something about the lowest signal that can be detected. It is usually defined as the input signal that produces a signal-to-noise ratio of one. This exposure is the noise equivalent exposure (NEE). The minimum signal is typically given as equivalent electrons rms. It is only one of many performance parameters used to describe system noise performance.

Figure 4-1 illustrates a signal transfer diagram. The shutter represents the integration time, t_{INT}. The detector converts the incident photons into electrons at a rate determined by the quantum efficiency. A voltage is created when the electrons are transferred to the sense node capacitance, C. The signal after the source follower amplifier is

$$V_{SIGNAL} = n_e \frac{Gq}{C} \qquad (4\text{-}1)$$

n_e is the total number of electrons in the charge packet. It includes both photoelectrons and dark current electrons. The device output conversion gain (OCG), Gq/C, has units of volts/electron with values typically ranging from 0.1 μV/e⁻ to 10 μV/e⁻. The source follower amplifier gain, G, is usually slightly less than one. Therefore, it is sometimes omitted from radiometric equations. Amplifiers, external to the CCD device, amplify the signal to a useful voltage. These amplifiers are said to be off-sensor or off-chip.

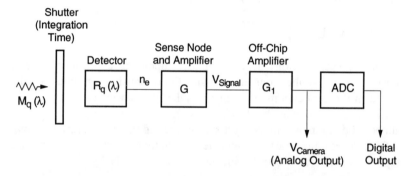

Figure 4-1. Representative signal transfer diagram. Many systems do not have a shutter, but it represents t_{INT}. The output may be specified as the number of electrons, output voltage of the source follower amplifier, or output of the off-chip amplifier. Some devices provide a digital output. V_{SIGNAL}, V_{CAMERA} or DN can be measured whereas n_e is calculated.

4.1.1. SPECTRAL RESPONSE

For the ideal material, when the photon energy is greater than the semiconductor band gap energy, each photon produces one electron-hole pair (quantum efficiency is one). However, the absorption coefficient is wavelength dependent and it decreases for longer wavelengths. This means that long wavelength photons are absorbed deeper into the substrate than short wavelengths. Very long wavelength photons may pass through the CCD and not be absorbed.

Any photon absorbed within the depletion will yield a quantum efficiency near unity. However, the depletion region size is finite and long wavelength photons will be absorbed within the bulk material. An electron generated in the substrate will experience a three-dimensional random walk until it recombines or reaches the edge of a depletion region where the electric field exists. If the diffusion length is zero, all electrons created within the bulk material will recombine immediately and the quantum efficiency approaches zero for these wavelengths. As the diffusion length approaches infinity, the electrons eventually reach a charge well and are stored. Here, the quantum efficiency approaches one. Doping controls the diffusion length and the quantum efficiency is somewhere between these two extremes (Figure 4-2). The quantum efficiency is dependent upon the gate voltage (low voltages produce small depletion regions) and the material thickness (long wavelength photons will pass through thin substrates).

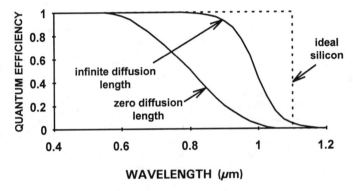

Figure 4-2. Theoretical internal quantum efficiency for silicon photosensors 250 µm thick (from reference 1). The long wavelength quantum efficiency depends upon the thickness of the substrate and the diffusion length. Most arrays have a relatively large diffusion length so that the quantum efficiency tends to be near the infinite diffusion length curve.

For front-sided illuminated devices using photogates, the spectral responsivity deviates from the ideal spectral response due to the polysilicon gate electrodes. The transmittance of polysilicon starts to decrease below 0.6 µm and becomes opaque at 0.4 µm (Figure 4-3). The blue region response can be increased by using thinner polysilicon gates. The gate structure is optically a thin film and therefore interference effects will cause variation in quantum efficiency that is wavelength dependent. Interference effects can be minimized through choice of material and film thickness. Since different manufacturing techniques are used for each array type, the spectral response varies with manufacturer. If a manufacturer uses different processes, then the spectral response will vary within his product line.

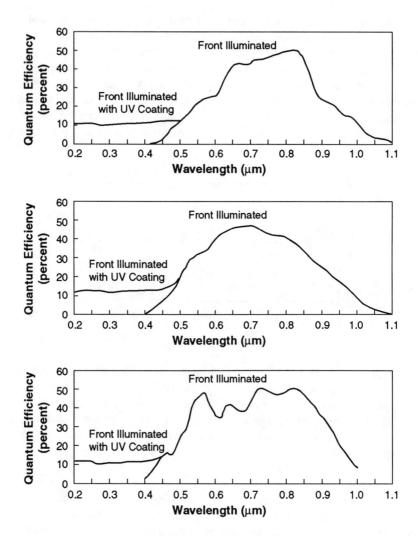

Figure 4-3. Typical response curves provided by different manufacturers. Above 0.8 µm, the quantum efficiency drops because the electron-hole pairs, generated deep in the array, may recombine before reaching a storage site. Below 0.6 µm, the polysilicon overcoat starts to become opaque. The variation in spectral response may not be a concern for many applications. Organic phosphors convert UV photons into visible photons and thereby extend the device's spectral response. The ripples in the responsivity are caused by interference effects.

ARRAY PERFORMANCE 95

Photodiodes, as used in most interline transfer devices, do not have the polysilicon gates. Therefore these devices tend to have a response that approaches the ideal silicon spectral response. That is, there are no interference effects and the blue region response is restored.

The UV response can be enhanced with a UV fluorescent phosphor. These phosphors are deposited directly onto the array and they emit light at approximately 0.54 to 058 μm when excited by 0.12 - 0.45 μm light. Phosphors radiate in all directions and only the light fluorescing toward the array is absorbed. Lumogen, which is one of many available organic phosphors, has an effective quantum efficiency of about 15%. It does not degrade quantum efficiency in the visible region because it is transparent at these wavelengths. The coating does not change the spectral response of the detector elements. Rather, it converts out-of-band photons into in-band photons. Therefore, the detector with the coating appears to have a spectral response from 0.12 to 1.1 μm.

Illuminating the array from the back side (BCCD) avoids the polysilicon problem and increases the quantum efficiency below 0.6 μm (Figure 4-4). Photons entering the back side are absorbed in the silicon and diffuse to the depletion region. However, short wavelength photons are absorbed near the surface and these electron-hole pairs recombine before reaching a storage site in a thick wafer. Therefore, the wafer is thinned to 10 μm or less to maintain good spectral responsivity (Figure 4-5). In back-sided thinned devices, the incident photon flux does not have to penetrate the polysilicon gate sandwich structure and interference effects are much easier to control. With a proper anti-reflection coating, a quantum efficiency of approximately 85% is possible. Owing to its extremely complex and fragile design, back-side illuminated devices are usually limited to scientific applications requiring high quantum efficiency.

CCD arrays for consumer and industrial applications are within a sealed environment for protection. Light must pass though a window (may be glass or quartz) to reach the array. In addition to reflection losses at all wavelengths, the glass transmittance decreases for wavelengths below 0.4 μm. For scientific applications requiring high sensitivity, the glass can be coated with a broad band anti-reflection coating. For UV applications, a quartz UV transmitting window can be used.

Arrays are cooled to reduce dark current and dark current shot noise (see Section 3.3., *Dark Current*, page 73). However, cooling also reduces the quantum efficiency with the red region of the spectrum affected the most.

96 CCD ARRAYS, CAMERAS, and DISPLAYS

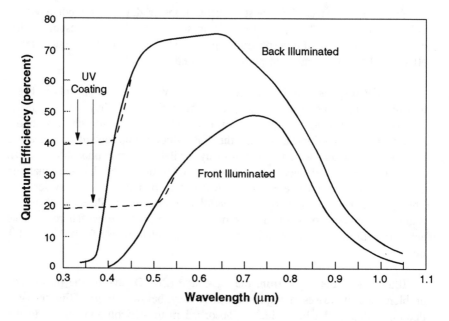

Figure 4-4. Representative spectral response of front-illuminated and back-side illuminated CCD arrays. The spectral response depends upon the process used to manufacture each device.

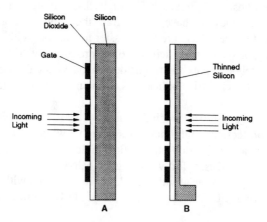

Figure 4-5. (a) Front illuminated and (b) back side illuminated arrays. The thinned back side illuminated array is used for scientific applications where high quantum efficiency is required.

4.1.2. RESPONSIVITY

The spectral quantum efficiency is important to the scientific and military communities. When the array is placed into a consumer video or industrial camera, it is convenient to specify the output as a function of incident flux density or energy density averaged over the spectral response of the array.

Assume that the array is at a distance of R_1 from an ideal source whose area is A_S. This radiometric setup is similar to that described in Section 2.1.3., *Camera Formula*, page 22, except no lens is present. For convenience, radiant quantities are used rather than photon flux. Assuming a Lambertian source,

$$n_{pe} = \frac{1}{q} \frac{A_S A_D}{R_1^2} \int_{\lambda_1}^{\lambda_2} L_e(\lambda) R_e(\lambda) t_{INT} \, d\lambda \qquad (4\text{-}2)$$

$R_e(\lambda)$ is the spectral response in amps/watt. The array output voltage (after the source follow amplifier) is $(G\,q\,n_{pe})/C$:

$$V_{SIGNAL} = \frac{G}{C} \frac{A_S A_D}{R_1^2} \int_{\lambda_1}^{\lambda_2} L_e(\lambda) R_e(\lambda) t_{INT} \, d\lambda \qquad (4\text{-}3)$$

It is desirable to express the responsivity in the form

$$V_{SIGNAL} = R_{AVE} \left(\frac{1}{A_D} \frac{A_S A_D}{R_1^2} \int_{\lambda_1}^{\lambda_2} L_e(\lambda) t_{INT} \, d\lambda \right) \qquad (4\text{-}4)$$

R_{AVE} is an average response that has units of volts/(Joule-cm^{-2}) and the quantity in the brackets has units of Joules/cm^2. Combining the two equations provides

$$R_{AVE} = \frac{G}{C} A_D \frac{\int_{\lambda_1}^{\lambda_2} L_e(\lambda) R_e(\lambda) d\lambda}{\int_{\lambda_1}^{\lambda_2} L_e(\lambda) d\lambda} \qquad \frac{volts}{Joules - cm^{-2}} \qquad (4\text{-}5)$$

R_{AVE} is the slope of the output/input transformation (Figure 4-6). These equations are valid over the region that the array output/input transformation is linear.

Experimentally, the most popular approach to create Figure 4-6 is to vary the exposure (e.g., the integration time). The irradiance can be changed by moving the source (varying R_1). It is imperative that the source color temperature remains constant. Assuming an incandescent bulb is used, reducing the bulb voltage will reduce the output but this changes the color temperature. Changing the source temperature changes the integrals and this leads to incompatible results (see Section 2.5., *Responsivity*, page 37). The source intensity can only be varied by inserting neutral density filters.

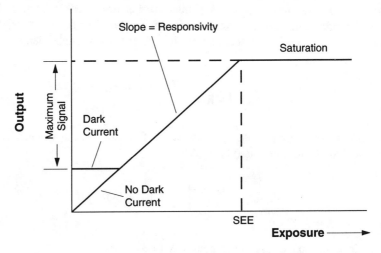

Figure 4-6. The average responsivity is the slope of the output-input transformation. The maximum input or the saturation equivalent exposure (SEE) is that input that fills the charge wells. SEE is used to define the dynamic range. Dark current limits the available signal strength. Cooling can reduce the dark current to a negligible level.

R_{AVE} is an average type responsivity that depends upon the source characteristics and the spectral quantum efficiency. While the source can be standardized (e.g., CIE illuminant A or illuminant D_{6500}), the spectral quantum efficiency varies by device (see Figure 4-3, page 94). Therefore extreme care must be exercised when comparing devices solely by the average responsivity.

Both $M_e(\lambda)$ and $R_e(\lambda)$ are functions of wavelength. If a very small wavelength increment is selected, $M_e(\lambda)$ and $R_e(\lambda)$ may be considered as constants. Equation 4-5 can be approximated as

$$R_{AVE} \approx \frac{G}{C} A_D \frac{M_e(\lambda_o) R_e(\lambda_o) \Delta\lambda}{M_e(\lambda_o) \Delta\lambda} \approx \frac{G}{C} A_D R_e(\lambda_o) \qquad (4\text{-}6)$$

The responsivity (in amps/watt) is related to the quantum efficiency R_q by $R_e = (q\lambda/hc) R_q$. If the wavelength is measured in microns, then $R_e = (\lambda/1.24) R_q$. Then

$$R_{AVE} \approx \frac{Gq}{C} \frac{\lambda_P}{hc} A_D R_q(\lambda_P) \quad \frac{volts}{Joules-cm^{-2}} \quad (4\text{-}7a)$$

or

$$R_{AVE} \approx \frac{G}{C} \frac{\lambda_P}{1.24} A_D R_q(\lambda_P) \quad \frac{volts}{Joules-cm^{-2}} \quad (4\text{-}7b)$$

$R_q(\lambda_P)$ is the peak quantum efficiency that occurs at λ_P. η is often used for the quantum efficiency: $R_{AVE} = (G \lambda_P A_D \eta)/(1.24\, C)$. This approximation is only valid for the specific wavelength selected.

If the device has a digital output, the number of counts (integer value) is

$$DN = int\left[V_{CAMERA} \frac{2^N}{V_{MAX}} \right] \quad (4\text{-}8)$$

N is the number of bits provided by the analog-to-digital converter. N is usually eight for consumer applications but may be as high as 16 for scientific cameras. V_{MAX} is the maximum output that corresponds to a full well, N_{WELL}:

$$V_{MAX} = G_1 \frac{Gq}{C} N_{WELL} \quad (4\text{-}9)$$

G_1 is the off-chip amplifier gain. The responsivity at the peak wavelength is approximated by

$$R_{AVE} \approx \left[\frac{2^N}{N_{WELL}} \frac{\lambda_P}{1.24} A_D R_q(\lambda_P) \right] \frac{DN}{joules/cm^2} \quad (4\text{-}10)$$

This assumes that the ADC input dynamic range exactly matches V_{MAX}. This may not be the case in real systems. If an anti-blooming drain is present, then the responsivity is defined up to the knee point (see Figure 3-28, page 79) and Equation 4-10 is modified accordingly.

Example 4-1
MAXIMUM OUTPUT

A device provides an output conversion gain (OCG) of 6 μV/e⁻. The well size is 70,000 electrons. What is the maximum output?

The maximum output is the OCG multiplied by the well capacity or 420 mV. If the dark current creates 5,000 electrons, then the maximum signal voltage is reduced to (70000-5000)(0.006 mV) = 390 mV. This signal may be amplified by an off-chip amplifier.

Example 4-2
DIGITAL RESPONSIVITY

A device with a well capacity of 50,000 electrons has its output digitized with a 12-bit analog-to-digital converter (ADC). What is the digital step-size (digital responsivity)?

The step size is the well capacity divided by the number of digital steps. $50000/2^{12}$ = 12.2 electrons/DN. Equation 4-2 (page 97) provides the number of electrons as a function of input exposure. The relationship between the ADC maximum value and charge well capacity assumes an exact match between V_{MAX} and the ADC input range. If these do not match, an additional amplifier must be inserted just before the ADC.

When using arrays for consumer video, the output is normalized to photometric units (see Section 2.2., *Photometry*, page 26). It is convenient to express the array average response as

$$V_{SIGNAL} = R_{PHOTOMETRIC} \left(683 \frac{1}{A_D} \int_{0.38}^{0.75 \, \mu m} M_p(\lambda) \, V(\lambda) \, d\lambda \right) \quad (4\text{-}11)$$

$R_{PHOTOMETRIC}$ has units of volts/lux and the bracketed term has units of lux. Combining Equations 4-3 and 4-11 yields

$$R_{PHOTOMETRIC} = \frac{\dfrac{G}{C}\dfrac{A_S A_D}{R_1^2} \int_{\lambda_1}^{\lambda_2} L_e(\lambda) R_e(\lambda) t_{INT}\, d\lambda}{683 \dfrac{1}{A_D} \int_{0.38}^{0.75\,\mu m} M_p(\lambda) V(\lambda)\, d\lambda} \quad \frac{volts}{lux} \quad (4\text{-}12)$$

$R_{PHOTOMETRIC}$ depends on the integration time and is either 1/60 sec or 1/30 sec depending on the architecture for EIA 170 or NTSC video standards (see Section 3.2.4., *Interline Transfer*, page 60). Experimentally, the photometric illuminance is measured with a calibrated sensor. The input intensity levels are changed either by moving the source (changing R_1) or by inserting neutral density filters between the source and the array.

The responsivity depends on the spectral response and the spectral content of the illumination. The apparent variation in output with different light sources was discussed in Section 2.5., *Responsivity*, page 37. Selecting an array based on the responsivity is appropriate if the anticipated scene illumination has the same color temperature as the calibration temperature. That is, if the photometric responsivity is measured with a CIE A illuminant and the scene color temperature is near 2856 K, then selecting an array with the highest photometric responsivity is appropriate. Otherwise, the average photometric responsivity is used for informational purposes only.

Example 4-3
PHOTOMETRIC OUTPUT

An array designed for video applications has a sensitivity of 250 μV/lux when measured with CIE illuminant A. What is the expected output if the input is 1,000 lux?

If the color temperature does not change and only the intensity changes, then the output is $(250\,\mu V)(1000) = 250$ mV. However, if the color temperature changes, the output changes in a nonlinear manner (See Section 2.5., *Responsivity*, page 37). Similarly, two sources may produce the same illuminance but V_{SIGNAL} may be quite different. The color temperature must always be specified when $R_{PHOTOMETRIC}$ is quoted.

Implicit in this calculation is that the integration time remains constant. Since photometric units are used for consumer video applications, it is reasonable to assume that t_{INT} is either 1/60 sec or 1/30 sec for EIA 170 compatibility.

4.1.3. MINIMUM SIGNAL

The noise equivalent exposure (NEE) is an excellent diagnostic tool for production testing to verify noise performance. NEE is a poor array-to-array comparison parameter and should be used cautiously when comparing arrays with different architecture. This is so because it depends on array spectral responsivity and noise. The NEE is the exposure that produces a signal-to-noise ratio of one. If the measured rms noise on the analog output is V_{NOISE}, then the NEE is calculated from the radiometric calibration.

$$NEE = \frac{V_{NOISE}}{R_{AVE}} \frac{joules}{cm^2} \, rms \qquad (4\text{-}13)$$

When referred back to the CCD array, NEE is simply the noise value in rms electrons. The absolute minimum noise level is the noise floor and this value is used most often for the NEE. Noise sources are discussed in Section 4.2., *Noise*. Although noise is a rms value, the notation, *rms*, is often omitted.

4.1.4. MAXIMUM SIGNAL

The maximum signal is that input signal that saturates the charge well and is called the saturation equivalent exposure (SEE). It varies with architecture, number of phases, and pixel size. The well size is approximately proportional to pixel area. Small pixels have small wells. If an anti-blooming drain is present, the maximum level is taken as the white clip level (see Figure 3-28, page 79).

For back-of-the-envelope calculations, the dark current is considered negligible (i.e., the device is cooled or MPP implemented). Then

$$SEE = \frac{V_{MAX}}{R_{AVE}} \frac{Joules}{cm^2} \qquad (4\text{-}14)$$

4.1.5. DYNAMIC RANGE

Dynamic range defined as the maximum signal (peak) divided by the rms noise. If an anti-blooming drain is present, the knee value is used as the maximum. Expressed as a ratio, the dynamic range is

$$DR = \frac{SEE}{NEE} = \frac{N_{WELL}}{\langle n_{SYS} \rangle} = \frac{V_{MAX}}{V_{NOISE}} \qquad (4\text{-}15)$$

$\langle n_{SYS} \rangle$ is the system noise measured in electrons rms. When expressed in decibels,

$$DR = 20 \log\left(\frac{SEE}{NEE}\right) \quad dB \qquad (4\text{-}16)$$

Sometimes the manufacturer will specify dynamic range by the peak signal divided by the peak-to-peak noise. This peak-to-peak value includes any noise spikes and, as such, may be more useful in determining the lowest detectable signal in some applications. If the noise is purely Gaussian, the peak-to-peak is assumed to be four to six times the rms value. This multiplicative factor is author dependent.

If the system noise is less than the analog-to-digital converter least significant bit (LSB), then the quantization noise limits the system dynamic range. The rms quantization noise, $\langle n_{ADC} \rangle$, is $V_{LSB}/\sqrt{12}$ where V_{LSB} is the voltage step corresponding to the least significant bit. The dynamic range is $2^N \sqrt{12}$. For a 12-bit ADC, the dynamic range is 14,189:1 or 83 dB. An 8-bit system cannot have a dynamic range greater than 887:1 or 59 dB.

☞
___

Example 4-4
DIGITAL DYNAMIC RANGE

A device with a well capacity of 50,000 electrons has a noise floor of 60 electrons rms. What size analog-to-digital converter should be used?

Adding noise powers, the total camera noise is related to the array noise and ADC noise by

$$\langle n_{SYS} \rangle = \sqrt{\langle n^2_{FLOOR} \rangle + \langle n^2_{ADC} \rangle} \qquad (4\text{-}17)$$

If an ADC is selected so that an LSB is equal to the array noise level, the ADC increases the camera noise by 4%. The array dynamic range is 50000/60 = 833 but the smallest available ADC has 10 bits or 1024 levels. This ADC devotes 1.23 bits to array noise. On the other hand, if the LSB was one-quarter of the array noise level, the camera noise would increase by only 1%. The ADC dynamic range must be greater than 3332:1. The smallest ADC that covers this range contains 12 bits.

Example 4-5
ARRAY PARAMETERS

An array has a dynamic range of 3000:1, average responsivity of 2 V/(μJ/cm^2), and a saturation equivalent exposure of 150 nJ/cm^2. These values apply only at λ_P. What are the remaining design parameters?

Maximum output:
    The maximum output is R_{AVE} SEE = 300 mV.
Noise level:
    The noise floor rms value is equal to the maximum output divided by the dynamic range or 100 μV rms.
Analog-to-digital Converter:
    For a digital output, the ADC must have 12 bits (2^{12} = 4096 levels) to span the array's full dynamic range. The responsivity in DN is 2^{12}/SEE = 27.3 DN/(nJ/cm^2) at λ_P.

4.2. NOISE

Many books and articles[3-9] have been written on noise sources. The level of detail used in noise modeling depends on the application. Shot noise is due to the discrete nature of electrons. It occurs when photoelectrons are created and when dark current electrons are present. Additional noise is added when reading the charge (reset noise) and introduced by the amplifier (1/f noise and white noise). If the output is digitized, the inclusion of quantization noise may be necessary.

Figure 4-1 (page 92) illustrated the signal transfer diagram. The noise transfer diagram is somewhat different (Figure 4-7). With different transfer functions, both the optical and electronic subsystems must be considered to maximize the system signal-to-noise ratio.

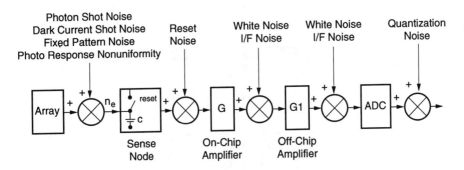

Figure 4-7. The various subsystems are considered as ideal elements with the noise introduced at appropriate locations.

Although the origin of the noise sources is different, they all appear as variations in the image intensity. Figure 4-8a illustrates the ideal output of the on-chip amplifier before CDS. The array is viewing a uniform source and the output of each pixel is identical. Photon shot noise produces a temporal variation in the output signal that is proportional to the square root of the signal level in electrons. Each pixel output will have a slightly different value (Figure 4-8b). Under ideal conditions, each pixel would have the same dark current and this value is subtracted from all pixels leaving the desired signal (see Section 3.3., *Dark Current*, page 73). However, dark current also exhibits fluctuations. Even after subtracting the average value, these fluctuations remain and create fixed pattern noise (Figure 4-8c). The output capacitor is reset after each charge is read. Errors in the reset voltage appear as output signal fluctuations (Figure 4-8d). The amplifier adds white noise (Figure 4-8e) and 1/f noise (Figure 4-8f). Amplifier noise affects both the active video and reset levels. The analog-to-digital converter introduces quantization noise (Figure 4-8g). Quantization noise is apparent after image reconstruction. Only photon shot noise and amplifier noise affect the amplitude of the signal. With the other noise sources listed, the signal amplitude, A, remains constant. However, the value of the output fluctuates with all the noise sources. Since this value is presented on the display, the processes appear as displayed noise. These noise patterns change from pixel-to-pixel and on the same pixel from frame-to-frame.

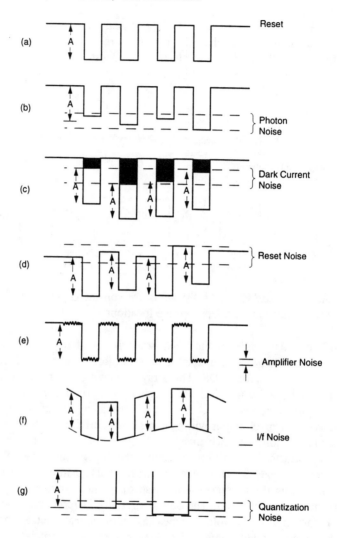

Figure 4-8. Noise sources affect the output differently. (a) though (f) represents the output after the on-chip amplifier. (g) is the output after the sample-and-hold circuitry. The array is viewing a uniform source that produces identical outputs (amplitude = A) for each pixel. (a) ideal output, (b) photon shot noise, (c), dark current shot noise, (d) reset noise, (e) amplifier noise, (f) amplifier 1/f noise, and (g) quantization noise. These values change from frame to frame. Pattern noise is not shown but is a variation that does not change significantly each frame. All these processes occur simultaneously.

Pattern noise refers to any spatial pattern that does not change significantly from frame-to-frame. In the dark, by pixel-to-pixel variations create fixed pattern noise (FPN). It is due to differences in detector size, doping density, and foreign matter getting trapped during fabrication. Photoresponse nonuniformity (PRNU) is the variation in pixel responsivity and is seen when the device is illuminated. This noise is due to differences in detector size, spectral response, and thickness in coatings. These "noises" are not noise in the usual sense. It occurs when each pixel has a different average value. This variation appears as spatial noise to the observer. Frame averaging will reduce all the noise sources except FPN and PRNU. Although FPN and PRNU are different, they are sometimes collectively called scene noise, pixel noise, pixel nonuniformity, or simply pattern noise.

It is customary to specify all noise sources in units of equivalent electrons at the detector output. Amplifier noise is reduced by the amplifier gain. When quoting noise levels, it is understood that the noise magnitude is the rms of the random process producing the noise. Noise powers are considered additive. Equivalently, the noise sources are RSSed (root sum of the squares) or added in quadrature:

$$\langle n_{SYS} \rangle = \sqrt{\langle n_{SHOT}^2 \rangle + \langle n_{PATTERN}^2 \rangle + \langle n_{RESET}^2 \rangle + \langle n_{ON-CHIP}^2 \rangle + \langle n_{OFF-CHIP}^2 \rangle + \langle n_{ADC}^2 \rangle}$$

(4-18)

$\langle n_i^2 \rangle$ is the noise variance for source i and $\sqrt{\langle n_i^2 \rangle} = \langle n_i \rangle$ is the standard deviation measured in rms units. Since photon shot noise and dark current shot noise are caused by the random generation of electrons, both contribute to shot noise: $\langle n_{SHOT}^2 \rangle = \langle n_{pe}^2 \rangle + \langle n_{DARK}^2 \rangle$.

Reset noise can be reduced to a negligible level with CDS (see Section 3.8., *Correlated Double Sampling*, page 83). Similarly, CDS can substantially reduce the source follower 1/f noise. The off-chip amplifier is usually a low-noise amplifier such that its noise is small compared to the on-chip amplifier noise. Although always present, quantization noise is reduced by selecting the appropriated sized analog-to-digital converter (e.g., selecting an ADC with many quantization levels).

For system analysis, it may be sufficient to treat the amplifier noise and other noise present as a single quantity and call it the noise floor. The array manufacturer usually provides this value and may call it readout noise, mux noise, noise equivalent electrons, or noise floor. It is the NEE (noise equivalent electrons) when all other noise sources have been removed. The value varies by

108 CCD ARRAYS, CAMERAS, and DISPLAYS

device and manufacturer. The simplified noise model is

$$\langle n_{SYS} \rangle \approx \sqrt{\langle n_{SHOT}^2 \rangle + \langle n_{FLOOR}^2 \rangle + \langle n_{PATTERN}^2 \rangle} \qquad (4\text{-}19)$$

The noise model does not include effects such as banding and streaking which are forms of pattern noise found in linear and TDI arrays. Banding can occur with arrays that have multiple outputs serviced by different nonlinear amplifiers (Figure 3-13, page 57). Streaking occurs when the average responsivity changes from column to column in TDI devices[10]. These effects are incorporated in the three-dimensional noise model (discussed in Section 12.2., *Three-dimensional Noise Model*) and are used to predict the performance of military systems. Only a complete signal-to-noise ratio analysis can determine which noise source dominates (discussed in Section 4.3., *Array Signal-to-noise Ratio*).

4.2.1. SHOT NOISE

Both photoelectrons and dark current contribute to shot noise. Using Poisson statistics, the variance is equal to the mean:

$$\langle n_{SHOT}^2 \rangle = \langle n_{pe}^2 \rangle + \langle n_{DARK}^2 \rangle = n_{pe} + n_{DARK} \qquad (4\text{-}20)$$

The number of photoelectrons was given by Equation 4-2, page 97, and the number of dark current electrons was given by Equation 3-5, page 74. These values should be modified by the CTE:

$$\langle n_{SHOT}^2 \rangle = \langle n_{pe}^2 \rangle + \langle n_{DARK}^2 \rangle = (CTE)^N n_{pe} + (CTE)^N n_{DARK} \qquad (4\text{-}21)$$

Since the CTE is high ($CTE^N \approx 1$), it is usually omitted from most equations. However, CTE should be included for very large arrays. With TDI, the number of photoelectrons and dark current increases with the number of TDI elements, N_{TDI}:

$$\langle n_{pe}^2 \rangle = N_{TDI} n_{pe} + N_{TDI} n_{DARK} \qquad (4\text{-}22)$$

While the dark current average value can be subtracted from the output to provide only the signal due to photoelectrons, the dark current shot noise cannot. Cooling the array can reduce the dark current to a negligible value and thereby reduce dark current shot noise to a negligible level (see Section 3.3., *Dark Current*, page 73).

4.2.2. RESET NOISE

The noise associated with resetting the sense node capacitor is often called kTC noise. This is due to thermal noise generated by the resistance, R, within the resetting FET. The Johnson noise current variance is

$$\langle i_n^2 \rangle = \frac{4kT}{R} \Delta f \qquad (4\text{-}23)$$

where T is the absolute temperature and k is Boltzmann's constant (k = 1.38 x 10^{-23} J/K). Since the resistance is in parallel with the sense node capacitor, the noise equivalent bandwidth is $\Delta f = RC/4$. Then

$$\langle i_n^2 \rangle = kTC \qquad (4\text{-}24)$$

When referred to the sensor, the rms noise in electrons is

$$\langle n_{RESET} \rangle = \frac{\sqrt{kTC}}{q} \quad e^- \text{ rms} \qquad (4\text{-}25)$$

$\langle n_{RESET} \rangle$ represents the uncertainty in the amount of charge remaining on the capacitor following reset.

For a 0.01 pF capacitor at room temperature, the noise is about 40 electrons rms and a 0.2 pf capacitor produces about 126 e⁻ rms. Reducing the capacitance reduces the noise. This has an added benefit that the device output conversion gain, Gq/C, increases. kTC noise may be significantly reduced with correlated double sampling. While cooling reduces this noise source, cooling is used primarily to reduce dark current. That is, cooling reduces dark current noise exponentially whereas reset noise is reduced only by $\sqrt{T}$.

4.2.3. ON-CHIP AMPLIFIER NOISE

Amplifier noise consists of two components: 1/f noise and white noise. If f_{KNEE} is the frequency at which the 1/f noise equals the white noise, then the amplifier noise density is

$$V_{ON-CHIP} = V_{AMP\,NOISE}\left(1 + \frac{f_{KNEE}}{f}\right) \quad \frac{rms\ volts}{\sqrt{Hz}} \quad (4\text{-}26)$$

1/f can be minimized through correlated double sampling.

When discussing individual noise sources, noise is usually normalized to unit bandwidth. When modified by electronic subsystems, the total system noise power is

$$\sigma^2_{SYS} = \int_0^\infty S(f_e)|H_{sys}(f_e)|^2\, df_e \quad (4\text{-}27)$$

$S(f_e)$ is the total noise power spectral density from all sources. $H_{sys}(f_e)$ is the frequency response of the system electronics. $H(f_e)$ is used by electronic circuitry designers and its magnitude is identical to the MTF. The noise equivalent bandwidth is that bandwidth with unity value that provides the same total noise power. Assuming the noise is white over the spectral region of interest $[S(f_e) = S_o]$, Δf_e is

$$\Delta f_e = \frac{\int_0^\infty S(f_e)|H_{sys}(f_e)|^2\, df_e}{S_o} \quad (4\text{-}28)$$

Δf_e applies only to those noise sources that are white and cannot be applied to 1/f noise. Although common usage has resulted in calling Δf_e the *noise bandwidth*, it is understood that it is a power equivalency.

If t_{CLOCK} is the time between pixels, the bandwidth for an *ideal* sampled-data system is

$$\Delta f_e = \frac{1}{2 t_{CLOCK}} = \frac{f_{CLOCK}}{2} \qquad (4\text{-}29)$$

Referred back to the array output,

$$\langle n_{ON\text{-}CHIP} \rangle = \frac{C}{Gq} V_{ON\text{-}CHIP\text{-}AMP\text{-}NOISE} \sqrt{\Delta f_e} \quad e^- \text{ rms} \qquad (4\text{-}30)$$

4.2.4. OFF-CHIP AMPLIFIER NOISE

The off-chip amplifier noise is identical in form to the on-chip amplifier noise (Equation 4-26). The 1/f knee value may be different for the two amplifiers. If f_{KNEE} is small, the noise referred to the array is

$$\langle n_{OFF\text{-}CHIP} \rangle = \frac{C}{GG_1 q} V_{OFF\text{-}CHIP\text{-}AMP\text{-}NOISE} \sqrt{\Delta f_e} \quad e^- \text{ rms} \qquad (4\text{-}31)$$

4.2.5. QUANTIZATION NOISE

The analog-to-digital converter produces discrete output levels. A range of analog inputs can produce the same output. This uncertainty, or error, produces an effective noise given by

$$\langle V_n \rangle = \frac{V_{LSB}}{\sqrt{12}} \quad e^- \text{ rms} \qquad (4\text{-}32)$$

V_{LSB} is the voltage associated with the least significant bit. For an ADC with N bits, $V_{LSB} = V_{MAX}/2^N$. When referred back to the array output

$$\langle n_{ADC} \rangle = \frac{C}{GG_1 q} \frac{LSB}{\sqrt{12}} \quad e^- \text{ rms} \qquad (4\text{-}33)$$

112 *CCD ARRAYS, CAMERAS, and DISPLAYS*

When the ADC is matched to the amplifier output, V_{MAX} corresponds to the full well. Then

$$\langle n_{ADC} \rangle = \frac{N_{WELL}}{2^N \sqrt{12}} \qquad (4\text{-}34)$$

Ideally, $\langle n_{ADC} \rangle$ is less than the noise floor. This is achieved by selecting a high resolution ADC (large N).

☞

Example 4-6
TOTAL NOISE

An array has a noise floor of 20 electrons rms, well capacity of 50,000 electrons and an 8-bit analog-to-digital converter. Neglecting shot noise and pattern noise, what is the total noise?

The LSB represents $50000/2^8 = 195$ electrons. The quantization noise is $195/\sqrt{12}$ or 56.3 electrons rms. The total noise is

$$\langle n_{SYS} \rangle = \sqrt{\langle n_{FLOOR}^2 \rangle + \langle n_{ADC}^2 \rangle} + \sqrt{20^2 + 56.3^2} \approx 59.7 \ e^- \ rms \qquad (4\text{-}35)$$

The array dynamic is range 2500:1 and the ADC only has 256 levels. Here, the ADC noise dominates the array noise. If a 12-bit ADC is used (4096 levels), the LSB represents 12.2 electrons and $\langle n_{ADC} \rangle = 3.5$ electrons rms. The total noise referred to the detector output is

$$\langle n_{SYS} \rangle = \sqrt{\langle n_{FLOOR}^2 \rangle + \langle n_{ADC}^2 \rangle} = \sqrt{20^2 + 3.52^2} \approx 20.3 \ e^- \ rms \qquad (4\text{-}36)$$

4.2.6. PATTERN NOISE

Fixed pattern noise (FPN) refers to the pixel-to-pixel variation[11,12] that occurs when the array is in the dark. It is primarily due to dark current differences. Significant FPN can be caused by synchronous timing generation effects at high data rates. It is a signal-independent noise and is additive to the other noise powers. Photoresponse nonuniformity (PRNU) is due to differences in responsivity (when light is applied). It is a signal-dependent noise and is a multiplicative factor of the photoelectron shot noise.

PRNU can be either be specified as a peak-to-peak value or a rms value referenced to an average value (may either be full well or one-half full well value). That is, the array is uniformly illuminated and a histogram of responses is created. The PRNU can either be the rms of the histogram divided by the average value or the peak-to-peak value divided by the average value. This definition varies by manufacturer so that the test conditions must be understood when comparing arrays. For this text, the rms value is used.

Since dark current becomes negligible by efficiently cooling the array, PRNU is the dominate pattern component for most arrays. As a multiplicative noise, PRNU is traditionally expressed as a fraction of the total number of charge carriers. If U is the fixed pattern ratio or nonuniformity, then

$$\langle n_{PATTERN} \rangle \approx \langle n_{PRNU} \rangle = U n_{pe} \qquad (4\text{-}37)$$

As with shot noise, CTE^N should be added to the equation. Since $CTE^N \approx 1$ for modest sized arrays, it is omitted from most equations. CTE should be included for very large arrays. TDI devices, though its inherent averaging, reduces PRNU by $1/N_{TDI}$. PRNU is usually supplied by the manufacturer. Off-chip gain/level correction algorithms can minimize FPN and PRNU. For system analysis, the corrected pattern noise value is used.

In principle, FPN and the noise floor can be reduced so that the system is photon shot noise limited. This provides the theoretical limit:

$$\langle n_{SYS} \rangle_{MIN} = \sqrt{n_{pe}} \qquad (4\text{-}38)$$

114 CCD ARRAYS, CAMERAS, and DISPLAYS

However, all systems have some pattern noise. PRNU is presented as a single multiplicative factor of the average photoelectron number. This approach assumes that the detectors are operating in a linear region with the only difference being responsivity differences. The plot of SNR versus nonuniformity suggests a desired maximum level of acceptable nonuniformity[13]. In Figure 4-9, the noise floor is considered negligible:

$$\langle n_{SYS} \rangle = \sqrt{n_{pe} + (U n_{pe})^2} \qquad (4\text{-}39)$$

$\langle n_{SYS} \rangle$ increases when PRNU is excessive. Typically an array is selected whose PRNU is inherently small. This value can be further reduced by off-chip electronics. The final value should be sufficiently small so that it does not contribute to the overall noise. Since cost increases as the PRNU decreases, the optimum PRNU occurs at the knee of the curve. This occurs when the photon shot noise is approximately equal to PRNU or $U \approx 1/\sqrt{n_{pe}}$. For worst case analysis, the charge well capacity should be used for n_{pe}.

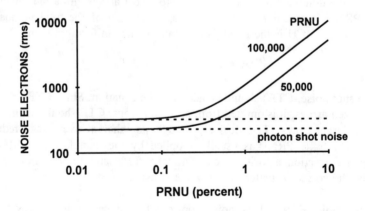

Figure 4-9. System noise as a function of PRNU for two signal levels. The noise floor is considered negligible. The noise floor will increase $\langle n_{SYS} \rangle$ and the location of the knee. $1/\sqrt{n_{pe}}$ is 0.4% and 0.3% when n_{pe} is 50,000 and 100,000 respectively. These values represent the "knee" in the curve.

4.2.7. PHOTON TRANSFER

Although various noise sources exist, for many applications it is sufficient to consider photon shot noise, noise floor, and PRNU. Here, the total array noise is

$$\langle n_{SYS} \rangle = \sqrt{\langle n_{SHOT}^2 \rangle + \langle n_{FLOOR}^2 \rangle + \langle n_{PRNU}^2 \rangle} \qquad (4\text{-}40)$$

or

$$\langle n_{SYS} \rangle = \sqrt{n_{pe} + \langle n_{FLOOR}^2 \rangle + (Un_{pe})^2} \qquad (4\text{-}41)$$

Either the rms noise or noise variance can be plotted as a function of signal level. The graphs are called the photon transfer curves and the mean-variance curve respectively. Both graphs convey the same information. They provide[14,15] array noise and saturation level from which the dynamic range can be calculated. As the charge well reaches saturation, electrons are more likely to spill into adjoining wells (blooming) and, if present, into the overflow drain. As a result, the number of noise electrons starts to decrease as the well reaches saturation (Figure 4-10).

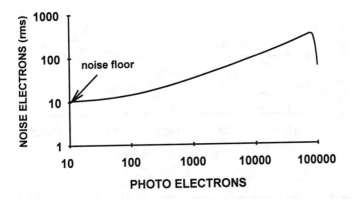

Figure 4-10. Photon transfer curve. At low input signals, the device noise is dominated by the noise floor. As the signal increases, the noise increases due to photon shot noise and pattern noise. As the well reaches saturation, there is a drop in noise electrons due to spill over and well capacity limitation.

For very low photon fluxes, the noise floor (or mux noise) dominates. As the incident flux increases, the photon shot noise dominates. Finally, for very high flux levels, the noise may be dominated by PRNU. As the signal approaches well saturation, the noise plateaus and then drops abruptly at saturation (Figure 4-10). Figure 4-11 illustrates the rms noise as a function of photoelectrons when the dynamic range is 60 dB. With large signals and small PRNU, the total noise is dominated by photon shot noise. When PRNU is large, the array noise is dominated by U at high signal levels.

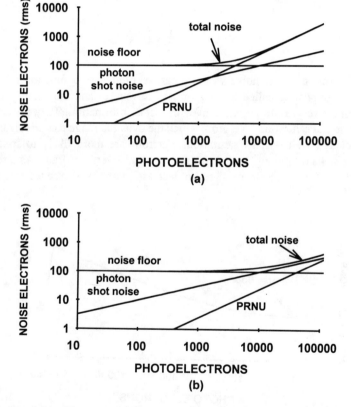

Figure 4-11. Photon transfer curves for (a) U = 2.5% and (b) U = 0.25%. The charge well capacity is 100,000 electrons and the noise floor (mux noise) is 100 e⁻ rms to produce a dynamic range of 60 dB. The slopes for noise floor, photon shot noise, and PRNU are 0, 0.5, and 1 respectively. Dark noise is considered negligible. The drop at saturation is not shown (see Figure 4-9).

Dark current shot noise only affects those applications where the signal-to-noise ratio is low (Figure 4-12). At high illumination levels, either photon shot noise or pattern noise dominate. Since many scientific applications operate in a low signal environment, cooling will improve performance. Consumer video and industrial cameras tend to operate in high signal environments and cooling will have little effect on performance. A full SNR analysis is required before selecting a cooled camera.

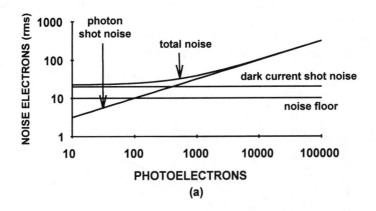

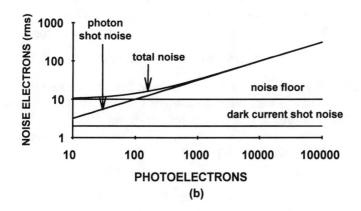

Figure 4-12. Dark current shot noise effects. The noise floor (mux noise) is 10 electrons rms and PRNU is zero. (a) Dark current shot noise is 20 e⁻ rms and (b) dark current shot noise is 2 e⁻ rms. For large signals, photon shot noise dominates the array noise. The drop at saturation is not shown (see Figure 4-9).

If both signal and noise are measured after the on-chip amplifier, then measured photo generated signal is $V_{SIGNAL} = G\,q\,n_{pe}/C$ and the measured noise is

$$V_{NOISE} = \frac{GC}{q}\langle n_{SYS}\rangle \qquad (4\text{-}42)$$

When the photon shot noise is one ($V_{NOISE} = 1$), the signal is the inverse of the output conversion gain: $V_{SIGNAL} = C/G\,q = 1/OCG$ (Figure 4-13). Similarly if the measurements are made after the off-chip amplifier, when the measured noise is one, $V_{SIGNAL} = C/G_1\,G\,q$. Finally if the values are collected after the analog-to-digital converter, when the shot noise is one, the shot noise signal provides the conversion in units of electrons/DN.

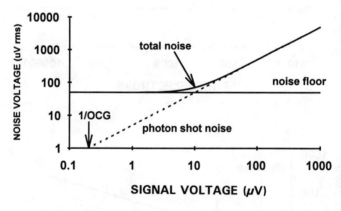

Figure 4-13. Photon transfer curve. The noise floor (mux noise) is 10 e⁻ and the OCG is 5 µV/e⁻. When the noise is one, the photon shot noise is 1/OCG. The drop at saturation is not shown (see Figure 4-9).

4.3. ARRAY SIGNAL-TO-NOISE RATIO

The array signal-to-noise ratio is

$$SNR = \frac{n_{pe}}{\sqrt{\langle n_{SHOT}^2 \rangle + \langle n_{FLOOR}^2 \rangle + \langle n_{PATTERN}^2 \rangle}} \quad (4\text{-}43)$$

With negligible dark current shot noise,

$$SNR = \frac{n_{pe}}{\sqrt{n_{pe} + \langle n_{FLOOR}^2 \rangle + (Un_{pe})^2}} \quad (4\text{-}44)$$

The charge transfer efficiency factor, CTE^N, should be added to n_{pe} in both the numerator and denominator. Since CTE^N is near one, it typically is omitted.

The SNR plots are similar to the photon transfer curves. At low flux levels, the SNR increases with the signal (noise floor is constant). At moderate levels, photon shot noise limits the SNR and at high levels, the SNR approaches 1/U. In Figure 4-14, the SNR is expressed as a power ratio which is $20\log(SNR)$. While the array may have a finite PRNU, it can be minimized off-chip through an appropriate algorithm.

When photon shot noise dominates, the SNR depends only on the photon flux level (SNR = $\sqrt{n_{pe}}$). The maximum available SNR is $\sqrt{N_{WELL}}$. The SNR is equal to the dynamic range only when the system is noise floor limited. This occurs only for low signal values and small wells (less than about 10,000 electrons). For large well capacity arrays, the actual SNR can never reach the value suggested by the dynamic range. Dynamic range is used to select an appropriate analog-to-digital converter resolution (number of bits). This assures that low contrast targets can be seen.

The array signal-to-noise ratio was derived for a large target that covers many pixels. The noise rides on the signal. The camera SNR, which includes all the noise sources, is provided in Section 6.1.2., *Camera SNR*. In Chapter 12, *Minimum Resolvable Contrast*, the average noise associated with the target and the background is considered.

120 CCD ARRAYS, CAMERAS, and DISPLAYS

When a target feature is described by a spatial frequency, f_o, the SNR is

$$SNR(f_o) = \frac{n_{pe} MTF_{SYS}(f_o)}{\sqrt{\langle n^2_{SHOT}\rangle + \langle n^2_{FLOOR}\rangle + \langle n^2_{PRNU}\rangle}} \qquad (4\text{-}45)$$

MTF_{SYS} is the subsystem MTFs that affect the signal. MTFs are discussed in Chapter 10, *System MTF*, and the application of Equation 4-45 is further discussed in Chapter 12, *Minimum Resolvable Contrast*.

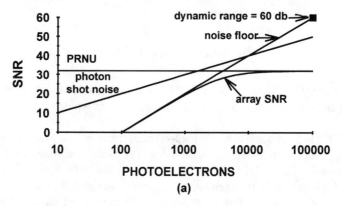

(a)

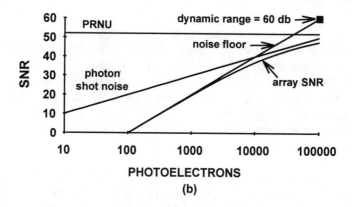

(b)

Figure 4-14. SNR as a function of noise for (a) U = 2.5% and (b) U = 0.25%. Dark current shot noise is considered negligible. For low signal levels, the SNR is limited by the noise floor. For high signals, the SNR is limited by photon shot noise when PRNU is negligible. The slopes of PRNU, photon shot noise, and noise floor (mux noise) are 0, 10, and 20 respectively.

4.4. REFERENCES

1. J. Janesick, "Introduction to Scientific Charge-Coupled Devices," Short Course SC-70 presented at the IS&T/SPIE's Symposium on Electronic Imaging: Science & Technology, San Jose CA (1995). J. Janesick presents this course at numerous symposia.
2. R. H. Dyck, "Design, Fabrication, and Performance of CCD Imagers," in *VLSI Electronics: Microstructure Science, Volume 3*, N. G. Einspruch, ed., pp. 70-71, Academic Press, New York (1982).
3. C. H. Sequin and M. F. Tompsett, *Charge Transfer Devices*, Academic Press, New York, NY (1975).
4. E. S. Yang, *Microelectronic Devices*, McGraw-Hill, NY (1988).
5. E. L. Dereniak and D. G. Crowe, *Optical Radiation Detectors*, John Wiley and Sons, New York, NY (1984).
6. D. G. Crowe, P. R. Norton, T. Limperis, and J. Mudar, "Detectors," in *Electro-Optical Components*, W. D. Rogatto, pp. 175-283. This is Volume 3 of *The Infrared & Electro-Optical Systems Handbook*, J. S. Accetta and D. L. Shumaker, eds., copublished by Environmental Research Institute of Michigan, Ann Arbor, MI and SPIE Press, Bellingham, WA (1993).
7. J. R. Janesick, T. Elliott, S. Collins, M. M. Blouke, and J. Freeman, "Scientific Charge-coupled Devices," *Optical Engineering*, Vol. 26(8), pp. 692-714 (1987).
8. T. W. McCurnin, L. C. Schooley, and G. R. Sims, "Charge-coupled Device Signal Processing Models and Comparisons," *Journal of Electronic Imaging*, Vol. 2(2), pp. 100-107 (1994).
9. M. D. Nelson, J. F. Johnson, and T. S. Lomheim, "General Noise Process in Hybrid Infrared Focal Plane Arrays," *Optical Engineering*, Vol. 30(11), pp. 1682-1700 (1991).
10. T. S. Lomheim and L. S. Kalman, "Analytical Modeling and Digital Simulation of Scanning Charge-Coupled Device Imaging Systems," in *Electro-Optical Displays*, M. A. Karim, ed., pp. 551-560, Marcel Dekker, New York (1992).
11. J. M. Mooney, "Effect of Spatial Noise on the Minimum Resolvable Temperature of a Staring Array," *Applied Optics*, Vol. 30(23), pp. 3324-3332, (1991).
12. J. M. Mooney, F. D. Shepherd, W. S. Ewing, J. E. Murguia, and J. Silverman, "Responsivity Nonuniformity Limited Performance of Infrared Staring Cameras," *Optical Engineering*, Vol. 28(11), pp. 1151-1161 (1989).
13. G. C. Holst, *Electro-Optical Imaging System Analysis*, pg. 360-362 and pg. 400-401, JCD Publishing, Winter Park, FL (1995).
14. J. E. Murguia, J. M. Mooney, and W. S. Ewing, "Diagnostics on a PtSi Infrared Imaging Array" in *Infrared Technology XIV*, I. Spiro, ed., SPIE Proceedings Vol. 972, pp. 15-25 (1988).
15. B. Stark, B. Nolting, H. Jahn, and K. Andert, "Method for Determining the Electron Number in Charge-coupled Measurement Devices," *Optical Engineering*, Vol. 31(4), pp. 852-856 (1992).

5
CAMERAS

CCD camera terminology is a mix of photographic, CCD chip, image processing, and television standards terminology. As CCD cameras enter emerging technologies, camera manufacturers adopt the marketplace terminology and add it to their specifications.

Many specifications include the known problems associated with vacuum image tubes such as image burn-in, distortion, or microphonics. These effects are not present in solid state devices. References to these vacuum tube disadvantages probably will disappear in the future as these tubes disappear.

Video standards were originally developed for televisions applications. The purpose of a standard is to insure that devices offered by different manufacturers will operate together. The standards apply only to the transmission of video signals. It is the camera manufacturer's responsibility to create an output signal that is compatible with the standards. Similarly, it is the display manufacturer's responsibility to insure that the display can recreate an adequate image.

Some agencies involved with standards include the Institute of Electrical and Electronic Engineers (IEEE), Electronics Industry Association (EIA), EIA-Japan (EIAJ), Joint Electron Devices Engineering Council (JEDEC), Society of Motion Picture and Television Engineers (SMPTE), International Electrotechnical Commission (IEC), Technical Center of the European Broadcast Union (EBU), Advanced Television Systems Committee (ATSC), Comite Consultatif International des Radiocommunications (CCIR) (now ITU-R), and International Telecommunications Union-Radiocommunications (ITU-R). Major manufacturers such as Panasonic, Sony, and JVC influence the standards.

There is no industry wide definition of "low-light-level" imaging system. To some, it is simply a CCD camera that can provide a usable image when the lighting conditions are about 1 lux. Lower illumination values may be possible by cooling the device. To others it refers to an intensified camera (ICCD). ICCDs are sometimes called low-light-level-television (LLLTV) systems.

Since CCD devices are designed for specific applications, the CCD array and camera use the same terminology: progressive scan chips are used in progressive scan cameras, TDI arrays are used in TDI cameras, etc. Through

CAMERAS 123

appropriate clocking, progressive scan cameras can also provide composite video and be used with consumer displays. Scientific cameras can have both analog and digital outputs. The corresponding sections in Chapter 3, *CCD Arrays*, should be read with the specific cameras listed here.

In this chapter, camera designs are separated into consumer/broadcast and industrial/scientific applications. The major features of each design are listed in the appropriate sections. However, as technology matures, the distinctions blur. That is, the same camera may be used for a variety of applications. New cameras offer variable frame rates and data rates that can be selected by the user. Therefore, the reader must review all sections before selecting a specific design. Chapter 6, *Camera Performance* discusses minimum signal, maximum signal, dynamic range, and signal-to-noise ratio.

Both monochrome and color cameras are quite complex as evidenced by the 1478-page book *Television Engineering Handbook*, K. B. Benson, ed., McGraw-Hill, New York (1986). This chapter is an overview of CCD camera operation. It highlights those features and specifications that are most often reported in camera data sheets. Together with the preceding chapters, it provides the necessary information to compare and analyze camera systems.

Selecting the right camera to meet specific requirements may be difficult. For consumer video, cost is very important. For broadcast systems, image quality and color are important. Scientific cameras offer low noise and high resolution. Issues not addressed include environmental conditions such as temperature, humidity, altitude, shock, vibration, the "-ilities" (e.g., reliability, maintainability etc.) and physical attributes (e.g., weight, size, and cost).

5.1. CAMERA OPERATION

Figure 5-1 illustrates a generic CCD camera. The lens assembly images the light onto the detector array. General purpose consumer cameras typically have an infrared blocking filter to limit the camera's spectral response to the visible spectral band (see Section 2.5., *Responsivity*, page 37). Depending upon the application (military, industrial, or scientific), the camera may or may not have the filter. Color cameras have a spectral response limited to the visible region.

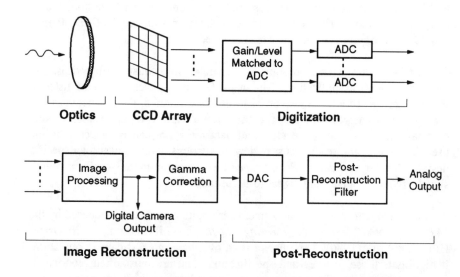

Figure 5-1. Generic functional block diagram. Not all components exist in every camera. The gamma correction circuitry is appropriate when the video is to be viewed on a CRT-based display.

A single chip color camera typically has an unequal number of red, green, and blue detectors resulting in different sampling frequencies for the different colors (discussed in Section 8.3.2., *Detector Array Nyquist Frequency*). Since the detectors are at discrete locations, target edges and the associated ambiguities are different for each color. A black-to-white edge will appear to have colors at the transition and monochrome periodic imagery may appear to have color. Aliasing is extremely annoying in color systems. A birefringent filter placed between the lens and the array reduces aliasing (discussed in Section 10.3.4., *Optical Anti-Alias Filter*). Any method to reduce aliasing generally reduces the overall modulation transfer function.

While aliasing is never considered desirable, it is accepted in scientific and military monochrome applications where the MTF must be maintained. Aliasing is reduced to an acceptable level in consumer color sensors. For specific applications such as medical imaging, it may be considered unacceptable. In medical imagery, an aliased signal may be misinterpreted as a medical abnormality requiring medication, hospitalization, or surgery.

CAMERAS 125

Correlated double sampling reduces reset noise and minimizes 1/f noise (see Section 3.8., *Correlated Double Sampling*, page 83). CDS circuitry may be integrated into the array package and this makes processing easier for the end user.

Image processing or image reconstruction reformats the array data into signals and timing that are consistent with transmission and display requirements. The advantage of a digital camera is that the output can be directly fed into an image processor, machine vision system, or digital data capture system. If using an analog output camera for machine vision, the video signal must pass through an external analog-to-digital converter. These multiple conversions may modify the imagery (discussed in Section 8.4., *Aliasing in Frame Grabbers and Displays*). Thus, staying within the digital domain ensures maximum image fidelity.

To present a linear transformation from a scene to an observer, a gamma correction (discussed in Section 5.3.3., *Gamma Compensation*) algorithm provides compensation for the monitor's nonlinear response. The concern with gamma should be limited to applications where cathode ray tube (CRT) based displays are used. As the display industry moves from CRTs to flat panel displays, gamma correction is not required. However, gamma provides some image enhancement and therefore will probably be used with flat-panel displays. Gamma correction is generally not preferred for machine vision or scientific applications.

The digital-to-analog converter (DAC) and sample-and-hold circuitry converts the digital data into an analog stair-step signal. The post-reconstruction filter removes the stair-step appearance and creates a smooth analog signal. After the post-reconstruction filter, the spatial scene sampling by the detectors is barely noticeable.

5.2. VIDEO FORMATS

In 1941, the Federal Trade Commission (FCC) adopted the 525-line, 30-Hz frame standard for broadcast television that was proposed by NTSC (National Television Systems Committee). The standard still exists today with only minor modifications. Although originally a monochrome standard, it has been modified for color transmission. Amazingly, color video fits within the video electronic bandwidth originally set aside for black-and-white television. NTSC initially selected a 60-Hz field rate. It was later changed to 59.94 Hz for color systems to avoid interactions (beat frequencies) between the audio and color subcarrier frequencies. The monochrome standard is often called EIA 170 (originally called RS 170) and the color format is simply known as NTSC (originally called EIA 170A or RS 170A). The timing for these two standards is nearly the same[1].

Worldwide, three color broadcast standards exist: NTSC, PAL, and SECAM. NTSC is used in the US, Canada, Central America, some of South America, and Japan. Phase alteration line system (PAL) was standardized in most European countries in 1967 and is also found in China. In 1967, France introduced SECAM (Sequentiel colour avec mémoire). SECAM is used in France, Russia, and the former Soviet bloc nations. Reference 2 provides a complete world wide listing. These standards are incompatible for a variety of reasons.

Video transmission bandwidth restrictions and the perception of good imagery on a display dictated the standard formats. That is, camera design is based upon the perception of displayed information. The interlace and scan times were selected to minimize electronic bandwidth and still provide a cosmetically pleasing image to the viewer with no flicker effects. Two fields are interwoven on the display. The eye's persistence time combined with the CRT phosphor persistence blends the two separate images into a perceived full image.

In the NTSC and EIA 170 standards, a frame is composed of an equivalent of 525 scan lines separated into two sequential fields (even and odd). Because time is required for vertical retrace, only 485 lines are displayed. There are 242.5 active lines per field. Since field 1 ends in a half-line, it is called the odd field. Figure 5-2 illustrates how the information is presented on a display. For commercial television, the aspect ratio (horizontal to vertical extent) is 4:3.

When a camera is designed to be compatible with a transmission standard, the manufacturer uses the broadcast terminology as part of the camera specifications. Similarly, consumer display manufacturers use the same terminology.

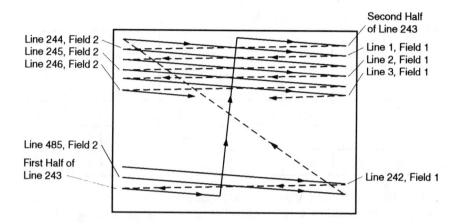

Figure 5-2. Presentation of EIA 170 video on a display. Standard timing is defined for video transmission only. CCD arrays may not read out data in this precise order. The camera image processing algorithm formats the data into a serial stream that is consistent with the video standard.

5.2.1. VIDEO TIMING

Table 5-1 provides the relevant parameters for the monochrome and color video standards. The standards specify the line frequency and blanking period. The scanning frequency is the frame rate multiplied by the number of lines. The total line time is calculated from the line frequency (e.g., 1/15750 = 63.492 μsec) and the active line time is the total line time minus the horizontal blanking period. The blanking period has a tolerance of about $\pm 0.25\%$ so that the active time line also varies slightly. Table 5-1 provides the minimum active time line where the maximum blanking period has been selected.

Table 5-1
STANDARD VIDEO TIMING

FORMAT	FRAME RATE (Hz)	LINES PER FRAME	ACTIVE LINES	SCANNING FREQUENCY (kHz)	TOTAL LINE TIME (μs)	MINIMUM ACTIVE LINE TIME (μs)
EIA 170	30	525	485	15.750	63.492	52.092
NTSC	29.97	525	485	15.7343	63.555	52.456
PAL	25	625	575	15.625	64.0	51.7
SECAM	25	625	575	15.625	64.0	51.7

For the monochrome video standard, the picture element was originally defined[3] as the smallest area being delineated by an electrical signal. The impulse response to an ideal band-limited system is

$$A(t) = \frac{\sin(2\pi f_c t)}{2\pi f_c t} \qquad (5\text{-}1)$$

f_c is the video bandwidth cutoff frequency. The pixel length is defined as the minimum distance between two impulses such that the two are still discernible. This occurs when the maximum of one falls on the first minimum (zero) of the next. The separation of these two vertical lines in time is $\Delta t = 1/2f_c$. This is the smallest possible dimension in a linear system. For the video standard, the number of pixels per line is

$$N = \frac{active\ line\ time}{\Delta t} = 2f_c(active\ line\ time) \qquad (5\text{-}2)$$

The nominal[4] video bandwidths of NTSC, PAL, and SECAM are 4.2, 5.5, and 6 MHz respectively. Table 5-2 lists the nominal number of horizontal elements per line. The quoted resolution (number of horizontal pixels) varies in the literature and depends upon the precise bandwidth and active line time selected by the author. Multiplying the pixel number by the number of active lines yields the elements per frame.

Table 5-2
NOMINAL NUMBER OF VIDEO PIXELS
Based upon electrical bandwidth only

FORMAT	NOMINAL BANDWIDTH (MHz)	ACTIVE LINE TIME (μs)	NUMBER of HORIZONTAL ELEMENTS	NUMBER of ACTIVE LINES	ELEMENTS PER FRAME
EIA 170	4.2	52.09	437	485	212,000
NTSC	4.2	52.45	440	485	213,000
PAL	5.5	51.7	569	575	327,000
SECAM	6.0	51.7	620	575	356,500

The full bandwidth is somewhat wider than the nominal value but high frequency components may be attenuated. For example, NTSC has a full bandwidth of approximately 4.5 MHz. The total bandwidth allocated for sound, chrominance, and luminance is 6, 8, and 8 MHz for NTSC, PAL and SECAM respectively. Using the allocated bandwidth, the maximum number of horizontal pixels is 625 for NTSC and 827 for PAL and SECAM.

Current camera design philosophy matches the number of detectors to the number of pixels supported by the video bandwidth. Based upon electrical bandwidth considerations, it appears that the number of horizontal detectors should range from 440 to 625 for NTSC, 560 to 827 for PAL, and 620 to 827 for SECAM compatible cameras.

The actual number of detector elements depends upon the manufacturer's philosophy. Image processing algorithm complexity is minimized when the spatial sampling frequency is equal in the horizontal and vertical directions. The 4:3 aspect ratio suggests that there should be 4/3 times more detectors in the horizontal direction compared to the vertical direction. Since the number of elements in the vertical direction is simply equal to the number of active scan lines, then the number of horizontal detectors should be about 646 for NTSC, and 766 for PAL and SECAM.

The number of pixels per line is often chosen so that the horizontal clock rate is an integer multiple of the color subcarrier. For NTSC, 768 pixels equates to four times the NTSC color subcarrier.

This larger number provides more sampling for each color and reduces color aliasing. For single chip color cameras, resolution is more difficult to define. The primaries or their complements spatially sample the scene unequally

130 CCD ARRAYS, CAMERAS, and DISPLAYS

(discussed in Section 8.3.2., *Detector Array Nyquist Frequency*). If the chip contains 768 detectors horizontally, then the number of green, red, and blue detectors is typically 384, 192, and 192 respectively. Since the eye is most sensitive to luminance values and the green component represents luminance, then the green resolution may be taken as the camera resolution. Most chips contain more green detectors and therefore the resolution is more than just one-third of an equivalent monochrome system. Through data interpolation, any number of output data points can be created. This number should not be confused with the camera spatial resolution. Single chip color cameras cannot provide the resolution listed in Table 5-2.

To reduce cost, a camera may contain fewer detector elements. Here, the full bandwidth of the video standard is not exploited. The imagery of these cameras will be poorer compared to those cameras that fully use the available bandwidth. Recall that the standard only specifies the timing requirements and a maximum bandwidth. The timing ensures that the camera output can be displayed on a standard display.

For the complete *system*, the horizontal resolution may be limited by the detector array, bandwidth of the video standard, or by the display. Display resolution is discussed in Section 7.5., *Resolution*. It is common for the display to limit the overall system resolution.

For NTSC and EIA 170 compatible arrays, it is common to about 480 detectors in the vertical direction. The image formatting algorithm inserts blank lines (all black) to meet video standards (485 lines). Since many displays operate in an overscan mode, the blank (black) lines are not noticed.

Resolution is often expressed in TV lines. This visual psychophysical measure was developed to specify the resolution of monitors. Because of phasing effects, the number of individual horizontal lines that can be perceived, on the average, is less than the number of scan lines. The ratio of the average number to the total number is the Kell factor[5] and is assumed to be 0.7. Thus, the vertical resolution is 340 lines for NTSC, and 402 lines for PAL and SECAM. This vertical resolution applies to the monitor capability and not the camera. However, the camera output must be displayed on a monitor so that the Kell factor is sometimes included in camera specifications.

For higher vertical resolution, more lines are required. EIA 343A (originally RS 343A) was created[6] for high resolution closed circuit television cameras (CCTV). Although the standard encompasses equipment that operates from 675 to 1023 lines, the recommended values are 675, 729, 875, 945, and 1023 lines per frame. 875 lines/frame is the most popular in the military (Table 5-3). Military systems require "standards conversion" boxes to record data and present imagery on EIA 170 compatible equipment. EIA 343A is still a monochrome standard.

Table 5-3
RECOMMENDED EIA 343A VIDEO FORMATS

LINES/FRAME	ACTIVE LINES	SCANNING FREQUENCY (kHz)	LINE TIME μs	ACTIVE LINE TIME μs
675	624	20.25	49.38	42.38
729	674	21.87	45.72	38.72
875	809	26.25	38.09	31.09
945	874	28.35	35.27	28.27
1023	946	30.69	32.58	25.58

5.2.2. COMPONENT/COMPOSITE SIGNALS

The term composite is used to denote a video signal that encodes luminance, color, and timing information onto a single channel. It can be transmitted on a single wire such as cable television. NTSC, PAL, and SECAM are composite video signals. The signal must be decoded to create an image.

Component analog video (CAV) carries each part on a separate wire. RGB, or their complementary values are carried on three separate cables. It is common for the G to be a composite line and carry the sync signals. Component videos do not require encoding and therefore there is less chance that there will be color or luminance cross coupling. This implies that component imagery will produce a higher quality image.

As equipment is miniaturized (e.g., hand-held video recorders) and resolution increased (e.g., HDTV), there seems to a proliferation of recommended standards[7]. Most standards deal with color encoding (chrominance encoding). The luminance channel is identical with the green channel and is labeled Y. The other signals are color and the exact composition depends upon the standard. SECAM uses the primaries whereas the other formats use the complements. The chrominance values are related through matrices but the

132 CCD ARRAYS, CAMERAS, and DISPLAYS

signals are not compatible due to different line frequencies and sync signals. Since the eye is most sensitive to luminance, the luminance channel has the widest bandwidth and the chrominance channels have smaller bandwidths.

Matrixing can convert the RGB into any color coordinate system. With the ease of digital implementation, multiple outputs are available from many cameras. That is, a camera may offer RGB, NTSC, and Y/C outputs to provide flexibility to the user.

5.2.3. IRE UNITS

The Institute of Radio Engineers (IRE) assigned 140 IRE units to the peak-to-peak composite signal (Figure 5-3). This provides 100 IRE units for the active image signal and 40 IRE units for blanking. They assumed that the video is one volt peak-to-peak when terminated by 75 Ω. Thus, each IRE unit is equal to $1/140 = 0.714$ V.

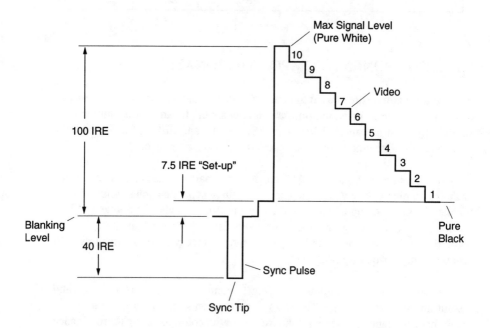

Figure 5-3. Definition of IRE units. The active video represents a gray scale divided into ten equal steps. NTSC assigns 7.5 IRE units to setup.

To avoid seeing the retrace, NTSC introduces a small setup. With most systems, the video is usually clamped at the blanking level and the video signal with setup does not contain a reference black signal. That is, the video does not contain a signal to indicate the black level. This voltage only occurs when there is a black object within the field-of-view. The blanking level (retrace) is sometimes called "blacker-than-black" since it is 7.5 IRE units below the black level. With the signal standardized to 1 V peak-to-peak, the setup reduces the active video to 0.66 volts.

By convention, the blanking level is assigned a value of zero volts so that the active video is positive while the sync pulse is negative. The peak-to-peak amplitude remains at 1 V. As new standards are introduced[7], the voltage levels associated with IRE units change (Table 5-4). The composite signal with setup (NTSC) can introduce ambiguity in signal level since there is no fixed black reference. Without setup, radiometrically correct imagery is ensured and black-to-white measurements can be performed with higher accuracy. SMPTE and EBU have recommended the non-NTSC format for color component video.

When a video signal is expressed as a percentage, it is numerically equal to the IRE units. 80% video is the same at 80 IRE units. For example, 80% video is $(0.8)(0.714) = 571$ mV with the NTSC standard.

Table 5-4
VIDEO STANDARDS

FORMAT	VOLTS			IRE UNITS			
	Sync tip	Setup	Peak	synch tip	blanking	setup	peak
NTSC	-0.286	0.0536	0.714	-40	0	7.5	100
Composite without setup	-0.286	-	0.714	-40	0	-	100
Non-NTSC	-0.300	-	0.700	-40	0	-	100

5.2.4. DIGITAL TELEVISION

Digital image processing is already present in most transmitters and receivers. It seems natural to transmit digitally. The major advantage is that digital signals are relatively immune to noise. Either a bit is present or it is not. A digital voltage signal, which may have been reduced in value and had noise added, can be restored to its full value with no noise in the receiver. An all-digital system avoids the multiple analog-to-digital and digital-to-analog conversions that degrade image quality.

Digital signal "transmission" was first introduced into tape recorders. Since a bit is either present or not, multiple generation copies retain high image quality. In comparison, analog recorders, such as VHS, provide very poor quality after just a few generations. The first digital recorder used a format called D-1 and it became known as the CCIR 601 component digital standard. Digital recording formats[8] now include D-1, D-2, D-3, D-4, and D-5.

The nominal 4.2-MHz bandwidth required by NTSC is not a continuous spectrum but consists of many bands due to the interaction of the line frequency (59.94 Hz) and scan frequency (15.7343 KHz). It is this discrete nature that allows the interleaving of chrominance, luminance, and audio within the allocated bandwidth. A comb filter, present in the receiver, separates the signals to minimize the cross talk that appears as aliased color. The composite video signal can be equated to a signal whose sampling frequency is 3.58 MHz for NTSC and 4.43 MHz for PAL. SECAM has different effective sampling rates for each color difference: 4.25 MHz and 4.41 MHz. These are the subcarrier frequencies[9].

The most popular sampling frequency is four times the subcarrier frequency and is simply called $4f_{sc}$ sampling. It is equal to 14.3 MHz for NTSC and 14.7 MHz for PAL. This allows the post-reconstruction filter to have a more gradual cutoff with less ripple. This filter is easier to design and is less expensive.
Since the data exists at discrete frequencies, strict adherence to the Nyquist criterion (discussed in Chapter 8, *Sampling Theory*) is not required.

In component video there is no subcarrier. A sampling frequency of 13.5 MHz has been selected to accommodate both 525 and 625 line formats. The luminance channel is sampled at 13.5 MHz and the two color difference channels are sampled at 6.75 MHz each. Recall that there is less detail in the color channels so they are sampled at a lower rate. If 13.5 MHz is four times the frequency of some imaginary subcarrier, then 6.75 MHz is twice that same imaginary frequency. This is known as 4:2:2 digital component television. If the three signals are multiplexed into a serial stream, the new sample rate is 27 MHz (13.5 + 6.75 + 6.75 = 27). In the multiplexed form, component signals can be transmitted on a single cable.

An image, whose amplitude is digitized into 8 bits, can provide a cosmetically acceptable image for consumer camcorders. However, non-linear circuitry is required to minimize the visibility of quantization contours at low signal levels. Otherwise, more bits are necessary. Photo quality imagery requires 12 or more bits. For composite signals, the sync pulse must also be digitized. The actual number of bits available to the imagery is conservatively less. CCIR 601 assigns 16 DN to the black level and 235 DN to the white level. The remaining values (1 to 15 and 236 to 254 DN) allow room for overshoots and control of the white and black levels. 0 and 255 DN are used for syncs.

5.2.5. HDTV

The current accepted meaning of high definition television (HDTV) is a television *system* providing approximately twice the horizontal and vertical resolution of present NTSC, PAL, and SECAM systems. While the term HDTV implies a receiver, the industry is equally concerned about transmission standards. The video format must be standardized before building cameras and receivers.

HDTV is envisioned as a system that provides high resolution on large displays and projection screens. Its resolution will not be fully appreciated in the typical living room where the sofa is nine feet from the television (discussed in Section 7.1., *The Observer*). The Japan Broadcasting Corporation (NHK) provided the first HDTV demonstration in 1981. The aspect ratio was initially 5:3 but this was later changed to 16:9 to match the aspect ratios used in motion pictures.

The goal is to have world wide compatibility so that HDTV receivers can display NTSC, PAL, and SECAM transmitted imagery. With today's multimedia approach, any new standard must be compatible with a variety of imaging systems ranging from 35-mm film to the various motion picture formats. The desire is to create composite imagery[10,11] and special effects from a mixture of film clips, motion pictures, and 35-mm film photos. Compatible frame rates allow the broadcasting of motion pictures and the filming of television.

As various companies entered the HDTV arena, they proposed different transmission formats[12]. To standardize the format, the Grand Alliance was created in May 1993. The seven members are: AT&T, David Sarnoff Research Center, General Instrument Corporation, MIT, North American Philips, Thomson Consumer Electronics, and Zenith Electronics. The Grand Alliance collaborates with a variety of agencies involved with standards (see list on page 122).

In April 1994, the Grand Alliance approved two different resolutions for HDTV (Table 5-5). Colorimetry is described in the SMPTE 240M standard and the 1080 x 1920 format is covered in the SMPTE 274M standard. Both formats have square pixels. The selected line rate should simplify conversion from NTSC and PAL into HDTV. 1125 is 15/7 times 525 and 9/5 times 625. Similarly, 787.5 is 3/2 times 525 but is approximately 100/79 times 625 lines. Thus, with the lower line rate, it is more difficult to convert PAL into HDTV.

Table 5-5
GRAND ALLIANCE RECOMMENDED FORMATS
(Recommended in April 1994)

RESOLUTION active lines x pixels/line	LINES PER FRAME	FRAME RATE (Hz)	SCAN TECHNIQUE
1080 x 1920	1125	24	Progressive
		30	Progressive
		60	Interlaced
720 x 1280	787.5	24	Progressive
		30	Progressive
		60	Progressive

Doubling the resolution, increasing the aspect ratio, and separating color from luminance requires an increased bandwidth of eight fold over current television requirements. With a nominal bandwidth of 4.2 MHz, the required HDTV bandwidth is approximately 33.6 MHz. A 10-bit system will have a data rate of 336 Mbit/sec. At this data rate, transmission losses in conventional wire cable limit the maximum transmission distance to, perhaps, 1 km. This demands image compression.

The FCC has allocated only a 6-MHz bandwidth for commercial television channels. Thus, image compression somewhere between 50:1 and 75:1 is required. The Grand Alliance recommends using MPEG-2 image compression. Although current cameras are analog devices[13], the Grand Alliance transmission format will be digital.

The final format has not been determined as of this writing. The FCC is expected to adopt a standard in 1996 and broadcasting should begin between 1998 and 2000. The status of HDTV cameras, standards, and receivers will appear in many journals over the next few years. Numerous articles on all aspects of television appear in the *SMPTE JOURNAL, IEEE Transactions on Broadcasting, IEEE Transactions on Consumer Electronic,* and *IEEE Transactions on Communications.*

5.3. CONSUMER/BROADCAST CAMERAS

Figure 5-4 illustrates a typical color camera functional block diagram. The CCD array output is processed to achieve the standard video signals. This section describes the knee, color correction, gamma, and aperture correction. Monochrome cameras tend to be used for scientific and industrial applications and are described in Section 5.4., *Industrial/Scientific Cameras.*

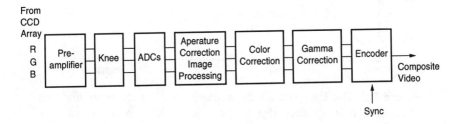

Figure 5-4. Typical color camera functional block diagram. Sync signals and other control logic are not shown. The subsystem location and design vary by manufacturer. Consumer cameras typically have a composite output that is created by the encoder. Scientific and industrial cameras may have a component output which is the signal before the encoder. To increase versatility, many cameras can provide either NTSC, RGB, or complementary color output. The RGB output provides easy interfacing with a computer monitor.

5.3.1. THE KNEE

The anti-blooming drain (see Section 3.5., *Anti-Blooming Drain*, page 78) creates a knee in the linear response to brightness. The knee slope and location can be any value (Figure 5-5). While scientific cameras operate in a linear fashion, there is no requirement to do so with consumer cameras. The consumer camera manufacturer will provide a mapping that he thinks will provide a cosmetically pleasing image. The knee also simplifies requirements in that an 8 bit ADC can provide aesthetically pleasing imagery over a wide input dynamic range.

Scientific and industrial applications often require a linear input-output system. For example, linearity is required for radiometric analysis. Interestingly, as some manufacturers add more knee control, others are actively designing algorithms to "de-knee" the signal before image processing.

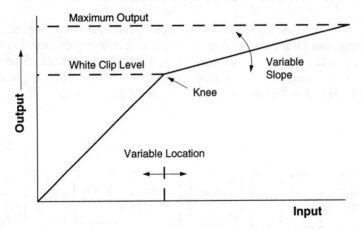

Figure 5-5. The knee extends the dynamic range by compressing high intensity signals. By deceasing the knee slope, the camera can handle a larger scene dynamic range.

5.3.2. COLOR CORRECTION

Figure 5-4 illustrated a typical color camera block diagram. The "color" signals may have originated from a single CFA (see Section 3.2.7., *Color Filter Arrays*, page 70) or three separate arrays as illustrated in Figure 5-6. Because of high cost, three-chip cameras were used almost exclusively for high-quality broadcast applications. Now, three-chip color cameras are entering the scientific market. Consumer camera designs, driven by cost, tend to use only one chip.

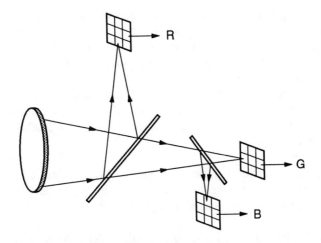

Figure 5-6. A high-quality color camera has three separate CCD arrays. One for each primary color or its complement. Light is spectrally separated by coatings on the beam splitters and additional filters. The three-chip camera is listed as CCDx3 or 3CCD in sales brochures. Each color samples the scene equally and this minimizes color Moiré patterns when compared to a CFA.

The color correction circuitry is an algorithm that can be represented as a matrix

$$\begin{pmatrix} R_2 \\ G_2 \\ B_2 \end{pmatrix} = \begin{pmatrix} k_{11} & k_{12} & k_{13} \\ k_{21} & k_{22} & k_{23} \\ k_{31} & k_{32} & k_{33} \end{pmatrix} \begin{pmatrix} R_1 \\ G_1 \\ B_1 \end{pmatrix} \qquad (5\text{-}3)$$

When used to improve color reproducibility, (i.e., matching the camera spectral response to the CIE standard observer), the process is called masking[14]. When manipulating signals into the NTSC standard, it is called matrixing.

Matrixing can be used to alter the spectral response and intensities of the primaries to compensate for changes in scene color temperature. Intensity changes are achieved simply by multiplying a primary by a constant. Spectral changes are approximated by adding various ratios of the primaries together. For example, if an object is viewed at twilight, the matrix can be adjusted so that the object appears to be in full sun light (neglecting shadows). A scene illuminated with a 2600 K source may appear somewhat yellow compared to a scene illuminated with a 9000 K source. This occurs because a 2600-K source has less blue than a 9000-K source.

Gamma correction is introduced into each of the three color channels. By convention, the analog signals are represented by E_G, E_R, and E_B. After gamma correction, they are represented by E_G', E_R', and E_B'.

In color filter arrays, like-colored sensitive detectors are separated resulting in a low fill-factor. The optical anti-alias filter blurs the image and effectively increases the optical area of the detector. The birefringent filter is placed between the lens assembly and the detector array. Birefringent crystals break a beam into two components: the ordinary and the extraordinary. This makes the detector appear larger optically. While this changes the MTF (discussed in Section 10.3.4, *Optical Anti-Alias Filter*), the sensitivity does not change. The light that is lost to adjoining areas is replaced by light refracted to the detector.

Sophisticated image processing is required to overcome the unequal sampling of the colors and to format the output into the NTSC standard. The algorithm selected depends upon the colored filter array architecture (see Figure 3-26, page 72). Algorithms may manipulate the ratios of blue-to-green and red-to-green[15], separate the signal into high and low frequency components[16], use a pixel shift processor[17], or add two alternating rows together[18,19] to minimize the colored edge artifact and false colors. The decoding algorithm is unique for each CFA configuration. Clearly, each CFA design requires a new decoding algorithm. Algorithm design is probably more of an art than a science since the acceptability of color aliasing is not known until after the camera is built.

Example 5-1
SINGLE COLOR CHIP ALGORITHM

Figure 5-7 illustrates a typical CFA[19] where rows are added to minimize aliasing. The NTSC odd field consists of CCD rows 1 + 2, 3 + 4, etc. and the NTSC even field consists of 2 + 3, 4 + 5, etc. Tables 5-6 and 5-7 provide the outputs.

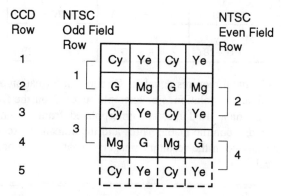

Figure 5-7. Typical CFA (From reference 20).

Table 5-6
SUMMING TWO ROWS

CCD ROW	OUTPUT LINE
1 + 2	1 (odd field)
3 + 4	3 (odd field)
2 + 3	2 (even field)
4 + 5	4 (even field)

Table 5-7
OUTPUT VALUES

TV LINE	PIXEL 1	PIXEL 2	PIXEL 3	PIXEL 4
1	Cy + G = 2G + B	Ye + Mg = 2R + G + B	Cy + G = 2G + B	Ye + Mg = 2R + G + B
3	Cy + Mg = R + G + 2B	Ye + G = R + 2G	Cy + Mg = R + G + 2B	Ye + G = R + 2G
2	Cy + G = 2G + B	Mg + Ye = 2R + G + B	Cy + G = 2G + B	Mg + Ye = 2R + G + B
4	Mg + Cy = R + G + 2B	Ye + G = R + 2G	Mg + Cy = R + G + 2B	Ye + G = R + 2G

The NTSC signal requires equal R, G, and B information on each line and this requires line interpolation. For example, pixel 3 on the first line does not contain any red data. Red must be interpolated from adjoining pixels. The selection depends upon the mathematics and the human eye response. The final selection depends upon the resultant imagery. That is, the final selection tends to be empirical.

5.3.3. GAMMA COMPENSATION

A consumer video camera may be purposely made nonlinear to match a monitor response. Gamma correction optimizes the displayed image when using CRT-based monitors. The input predistortion (created at the camera) can only be effective over a strictly defined range. The overall system is linear only when the correct display unit is used. Gamma correction is nonlinear and should be inserted after any required image processing. As flat panel displays replace CRT-based displays, the concern with gamma correction should disappear.

The luminous output of a cathode ray tube is related to its grid voltage by a power law

$$L_{DISPLAY} = K(V_{GRID})^\gamma \qquad (5-4)$$

If the camera has the inverse gamma (Figure 5-8), then the radiometric fidelity is preserved at the display.

$$L_{DISPLAY} = K(V_{GRID})^\gamma = K\left(V_{SCENE}^{\frac{1}{\gamma}}\right)^\gamma = K V_{SCENE} \qquad (5\text{-}5)$$

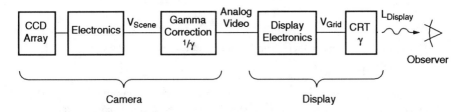

Figure 5-8. Gamma correction should only be used when the imagery is directly presented on a CRT-based display.

Although gamma correction circuitry is placed in the camera and it is numerically equal to the inverse of assumed CRT gamma, popular parlance simply calls it gamma. From now on, gamma will mean the inverse of the CRT gamma value. Gamma compresses small signals and some black detail is lost. In certain high contrast scenes, such as scenes with direct sunlight and shaded areas, the darker parts will lack contrast. It also changes the noise characteristics depending upon the scene level.

Figure 5-9 illustrates a machine vision system or where the imagery is evaluated by a computer. Gamma correction should be inserted before any image processing. Although standards recommend a gamma correction value, each manufacturer modifies these values to create what he perceives to be a cosmetically pleasing image. The output of an imaging processing work station is shown on a computer monitor. Although a machine vision system does not require a monitor, it adds convenience. A gamma correction, internal to the computer, assures that the correct image is shown on the computer monitor.

Although CRT displays have a range of gammas, NTSC, PAL, and SECAM "standardized" the display gamma to 2.2. 2.8, and 2.8 respectively. However, all three systems use 0.45 as the inverse. For NTSC, the monitor luminance is proportional to the detector output and target intensity (2.2)(0.45) = 1. With PAL and SECAM, the luminance still has a residual gamma[20] of 1.27. This provides some contrast enhancement.

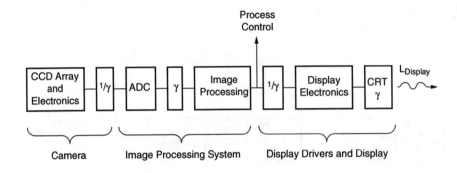

Figure 5-9. Image processing is simplified when gamma has been removed. The camera gamma should be one if the image is to be digitized and displayed on a computer monitor. Although not required for machine vision applications, the monitor offers convenience.

The gamma slope becomes very large for low values of the input brightness. It has therefore become accepted practice to limit the operation of the gamma compensator to an arbitrary level. SMPTE Standard 240M[21] was developed for high definition televisions but is applied to many systems It has since been replaced by ITU 709. Normalized to one, 240M states

$$V_{VIDEO} = 1.1115(V_{SCENE})^{0.45} - 0.1115 \quad 0.0228 \leq V_{SCENE} \leq 1 \quad (5\text{-}6a)$$

$$V_{VIDEO} = 4 V_{SCENE} \quad V_{SCENE} < 0.0228 \quad (5\text{-}6b)$$

V_{SCENE} is the voltage before the gamma corrector and V_{VIDEO} is the voltage after. $V_{SCENE} = 0.0228$ is the toe break intensity and the toe slope is 4 (when $L_{in} < 0.0228$). For gamma correction to be accurate, the black reference level must be known. If gamma is inserted after the 7.5% setup, some ambiguity exists in the black level resulting in some intensity distortion. While gamma correction may contain errors, the human eye is very accommodating. The consumer can also adjust the receiver contrast to optimize the image in accordance with his personal preferences. However, this luxury is not available to scientific applications.

Gamma compensation is often performed digitally using a look-up table. Figure 5-10 illustrates the desired output-input transformation recommended by SMPTE 240M. Alternate look-up tables are possible[22]. In fact, some camera manufacturers[19] developed an alternate gamma function *to enhance gray level in a high light intensity region and to obtain high S/N ratio in low light intensity regions simultaneously.* Changing the gamma to minimize the compression of dark areas is also called black stretching. Cameras may have a continuous variable gamma from 0.4 to 1 or may be preselected to 0.45 or 1. The user selects the appropriate value.

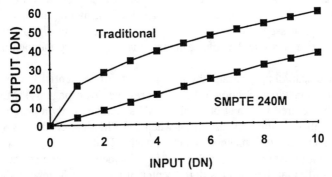

Figure 5-10. Gamma output-input transformation recommended by SMPTE 240M. Expansion of the first 10 levels of an 8-bit ADC system. Data only exists at integer values. Compression of blacks is obvious.

5.3.4. "APERTURE" CORRECTION

Image sharpness is related to the modulation transfer function (discussed in Chapter 9, *MTF Theory* and Chapter 10, *System MTF*). By increasing the MTF at high spatial frequencies, the image usually appears "sharper."

For analysis purposes, a camera is separated into a series of subsystems: optics, detector, electronics, etc. MTF degradation in any one subsystem will soften imagery. The degradation can be partially compensated with electronic boost filters (discussed in Section 10.5.4., *Boost*). Excessive boost creates ringing in the imagery which appears as ripples on target edges.

Historically, each subsystem was called an "aperture." Therefore boost provides "aperture correction." This should not be confused with the optical system where the optical diameter is the entrance aperture.

Boost filters also amplify noise so that the signal-to-noise ratio remains constant. However, for noisy images, the advantages of boost are less obvious. For systems that are contrast limited, boost may improve image quality. As a result, these filters are used only in high contrast situations (typical of consumer applications) and are not used in scientific applications where low signal-to-noise situations are often encountered.

5.4. INDUSTRIAL/SCIENTIFIC CAMERAS

The operation of industrial and scientific cameras is straight forward. They are built around the CCD device capabilities. The CCD array output is amplified and then either processed into a standard video format or left in digital format. Although these cameras may be color cameras, the majority are monochrome.

Progressive scan simply means the noninterlaced or sequential line-by-line scanning of the image (see Section 3.2.5., *Progressive Scan*, page 64). Moving objects can be captured with a strobe light or a fast shutter. Only a progressive scan camera will give the full resolution when the illumination time is short. These cameras can capture imagery asynchronously. That is, the camera timing is initiated from an external sync pulse that is coincident with the strobe flash which may signify that the object is in the center of the field-of-view. Although the image is captured in a progressive scan manner, the camera output is processed into a standard video output so that it can be displayed on commercial monitors. For critical applications, a digital output is available.

TDI offers another method to capture objects moving at a constant velocity. The operation of TDI arrays (see Section 3.2.6, *Time-Delay and Integration*, page 66) exactly describes how a TDI camera will operate. The TDI camera output may not be in a standard format and may need a dedicated computer or image processing work station. Reference 23 describes a TDI system used for remote sensing. Here, the aircraft or satellite provides the motion and the objects on the ground are stationary.

Consumer cameras are specified by the minimum detectable signal and dynamic range. Besides these metrics, scientific and industrial cameras are specified by MTF and the signal-to-noise ratio. Scientific cameras often have a large dynamic range and may have 10-, 12-, 14-, or 16-bit analog-to-digital converters. Key to the operation is that the ADC is linear.

Example 5-2
TDI CAMERA

Bottles move on a conveyor belt at the rate of 1200 feet/minute. An image processing system must detect defects without any human intervention. The bottle must be scrapped if a defect is larger than 0.1 inches. The camera sits 48 inches from the conveyor belt. If the detectors are 20 μm square, what focal length lens should be used? What is the camera line rate? If the bottle is 10 inches tall, what should the array size be?

Imaging processing software accuracy increases as the number of pixels on the target increases. While one pixel on the target may be sufficient for special cases, signal-to-noise ratio considerations and phasing effects typically require that the object cover at least three pixels. Thus, the image of the defect must be at least 60 μm.

The optical magnification is

$$M_{OPTICS} \frac{R_2}{R_1} = \frac{y_2}{y_1} = \frac{60 \, \mu m}{(0.1 \, inch)(25400 \, \mu m/inch)} = 0.0236 \quad (5\text{-}7)$$

R_1 is the distance from the lens to the object and R_2 is the lens to CCD array distance. The focal length is

$$fl = R_1 \frac{y_2}{y_1} = \frac{R_2 R_1}{R_2 + R_1} = 1.1 \, inches \quad (5\text{-}8)$$

The image is moving at (1200 ft/min)(1 min/60 sec)(304800 μm/ft)(0.0236) = 143865 μm/sec. Since the detectors are 20 μm, the line rate is 7193 lines/sec. This creates a detector dwell time of 1/7193 = 140 μsec. That is, the edge of the defect has moved across the detector element in 140 μsec. The pixel clock must operate at 7193 Hz and the integration time is 140 μsec.

The image of the bottle is (10 inches)(25400 μm/inch)(0.0236) = 5994 μm. With 20 μm square detectors, the array must contain 300 detectors in the vertical direction. The number of TDI elements depends upon the detector noise level, light level, and detector integration time. The signal-to-noise ratio increases by $\sqrt{N_{TDI}}$.

148 *CCD ARRAYS, CAMERAS, and DISPLAYS*

5.4.1. ANALOG-to-DIGITAL CONVERTERS

An analog-to-digital converter (ADC) is an essential component of all digital CCD cameras. For consumer applications, an 8-bit converter is used. Sometimes seemingly slight deviations from perfection in an ADC can modify image fidelity[24] when examined by a trained photo interpreter or by an automatic target detection algorithm. While a nonlinear ADC may not be visually obvious with a single readout, it may become obvious if the array has multiple outputs with each output having its own ADC. Here, any object which spans the two areas with different ADCs will have a discontinuity in intensity.

ADC linearity is specified by differential nonlinearity (DNL) and integral nonlinearity (INL) (Figure 5-11). The differential nonlinearity is a test of missing codes or misplaced codes. Two adjacent codes should be exactly one LSB apart. Any positive or negative deviation from this is the DNL. If the DNL exceeds one LSB, there is a missing code. Many nonlinearity specifications simply state "no missing codes." This means that the DNL is less than one LSB. However, 1.01 LSB DNL (fail specification) and 0.99 LSB DNL (pass specification) probably look the same visually. Poor DNL can affect the cosmetic quality of images.

Integral nonlinearity is the deviation of the transfer function from an ideal straight line. Most ADCs are specified with endpoint INL. That is, INL is specified in terms of the deviation from a straight line between the end points. Good INL is important for radiometric applications.

ADCs exist in most cameras. With commercially available cameras, the manufacturer has selected the ADC. To ensure that the ADC is linear, the user must experimentally characterize its performance in a manner consistent with its use[25]. This is rather difficult if the camera output is an analog signal. The analog output represents a smoothed output of the ADC. The smoothing can hide ADC nonlinearities.

The same problem may exist within a frame capture board. The user must test the board for any nonlinearities. Since the frame capture board usually resides within a computer, testing is relatively easy.

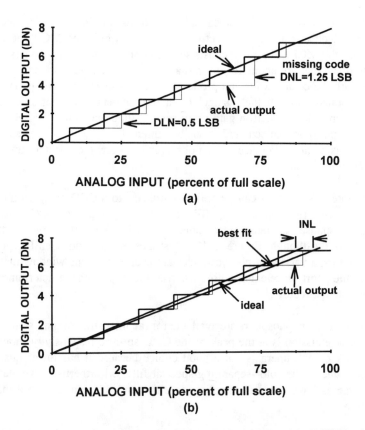

Figure 5-11. ADC nonlinearity for 3-bit ADC. (a) Differential nonlinearity and (b) integral nonlinearity. Missing codes can be caused by the accumulation of DNLs. Accumulation can also cause INL.

5.4.2. INTENSIFIED CCDs

Although low-light-level televisions can be used for many applications, they tend to be used for military and scientific applications. The term *image intensifier* refers to a series of special imaging tubes[26] that have been designed to internally amplify the number of photons. These tubes increase the number of incident photons by several orders of magnitude. The more popular second generation (GenII) and third generation (GenIII) intensifier tubes use a microchannel plate (MCP) for intensification. Third generation is similar in design to second generation with the primary difference that they have higher quantum efficiency.

When photons hit the photocathode, electrons are liberated and thereby create an electron image. The electrons are focused onto the MCP and about 80% enter the MCP. The MCP consists of thousands of parallel channels (glass hollow tubes) that are about 10 μm in diameter. Electrons entering the channel collide with the coated walls and produce secondary electrons. These electrons are accelerated through the channel by a high potential gradient. Repeated collisions provide an electron gain up to several thousand. Emerging electrons strike the phosphor screen and a visible intensified image is created. The intensifier is an analog device and an observer can view the resultant image on the phosphor screen.

The intensifier output can be optically linked[27] to a CCD array with either a relay lens or a fiber-optic bundle (Figure 5-12). The intensified CCD (ICCD) spectral response is determined solely by the intensifier photocathode responsivity. Although Figure 5-13 illustrates GenII and GenIII spectral responsivities, a variety of photocathodes are available[28]. The window material controls the short wavelength cutoff whereas the photocathode determines the long wavelength cutoff.

Although many phosphors are available for the intensifier output, a phosphor whose peak emission is at the peak of the CCD spectral responsivity should be selected. A color camera is not a good choice because it has lower quantum efficiency due to the color-separating spectral filters. Furthermore, single chip color cameras have unequal sampling densities and this can lead to a loss in resolution.

The ICCD can provide high-quality, real-time video when the illumination is greater than 10^{-3} lux (corresponding to starlight or brighter). For comparison, standard cameras (non-intensified) are sensitive down to about 1 lux. Thus, ICCDs offer high-quality imagery over an additional three orders of magnitude in illumination. A three-order magnitude increase in sensitivity implies that the MCP gain should be about a thousand. However, it must be an order of magnitude larger to overcome the transmission loss in the fiber optic bundle or relay lens. For very low illumination levels, it may be necessary to increase the integration time and then real-time video is no longer possible.

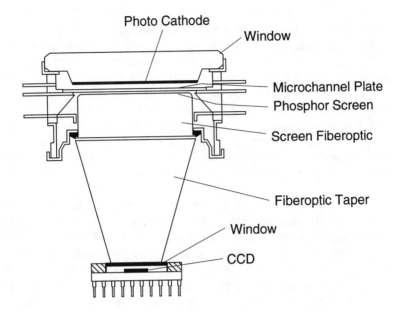

Figure 5-12. CCDs can be coupled to an image intensifier using either fiber-optic bundle or relay lens. The tapered fiber-optic bundle (shown here) tends to be the largest component of the ICCD.

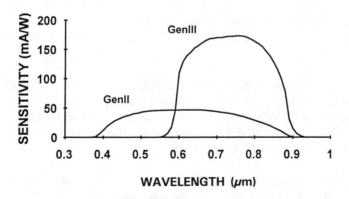

Figure 5-13. Nominal spectral response of militarized (fielded) GenII and GenIII image intensifier tubes. GenIII provides a better match to the night sky illumination (see Figure 2-13, page 36). The responsivity depends upon the window and photocathode material. GenII tubes use a multialkali photocathode whereas GenIII tubes use GaAs photocathodes.

By simply changing the voltage on the image intensifier, the gain can be varied. Thus, operation over a wide range of luminance levels ranging from starlight to daylight is possible. Gating is probably the biggest advantage of the ICCD. Electronic shutter action is produced by pulsing the MCP voltage. Gating allows the detection of low level signals in the presence of interfering light sources of much greater energy by temporal discrimination. Gate times as short as 5 ns FWHM (full width at half maximum) are possible. This allows the detection of laser diagnostic pulses against very intense continuous sources. Since full frame transfer or frame transfer CCDs are used for ICCD applications, the inherent gating capability of the intensifier is used as a shutter during transfer time. This eliminates image smear.

Since the CCD detector size is typically smaller than the intensifier's resolution, the intensifier's image must be minified. Fiber-optic bundles reduce camera size and weight (lenses are relatively heavy). But fiber-optic coupling is complex since it requires critical alignment of the fibers to the CCD pixels. The fiber area and pitch, ideally, should match the CCD detector size and pitch for maximum sensitivity and resolution. However, to alleviate critical alignment issues, the fiber diameter is typically one-half the pixel pitch. This reduces sensitivity somewhat.

There are four spatial samplers in the ICCD: (1), sampling by the discrete channels in the MCP, (2) the fiber-optic output window within the intensifier, (3) the input to the fiber-optic minifier, and (4) the CCD array. At these interfaces, the image is re-sampled. Depending upon the location (alignment) of the fiber-optic minifier, Moiré patterns can develop and the MTF can degrade. In addition, manufacturing defects in the fiber-optic can appear[29] as a "chicken wire" effect.

The lens coupled system avoids the alignment problems associated with fibers and does not introduce Moiré patterns. It offers more flexibility in that an image intensifier with an output lens can be connected to an existing CCD camera. The lens coupled ICCD represents a cost-effective way to add gating capability to an existing CCD camera.

The intensifier tube typically limits the ICCD resolution. Increasing the CCD resolution may have little effect on the system resolution. The CCD array is typically rectangular whereas the intensifier is round and the CCD cannot sense the entire amplified scene. Most commercially available ICCDs are based on the standard military 18-mm image intensifier. Larger tubes (25 mm and 40 mm) are available but are more expensive and increase coupling complexity.

Intensifiers are electron tubes. Tubes can bloom so that scenes with bright lights are difficult to image. For example, it is difficult to image shadows in a parking lot when bright street lights are in the field-of-view. Intensifiers are subject to the same precautions that should be observed when using most photosensitive coated pickup tubes. Prolonged exposure to point sources of bright light may result in permanent image retention.

The intensifier specifications alone do not accurately portray how the system will work. Only a detailed analysis will predict performance. Resolution of the CCD array will not describe system resolution if the image intensifier provides the limiting resolution. While the CCD photo response may be quite uniform, the overall nonuniformity is limited by the image intensifier. Standard military image intensifier photocathodes have nonuniformities approaching $\pm 30\%$. Only carefully selected intensifiers with low nonuniformity are used as part of the ICCD.

5.5. REFERENCES

1. The timing requirements can be found in many books. See, for example, D. H. Pritchard, "Standards and Recommended Practices," in *Television Engineering Handbook*, K. B. Benson, ed., pp. 21.40-21.57, McGraw-Hill, New York, NY (1986).
2. D. H. Pritchard, "Standards and Recommended Practices," in *Television Engineering Handbook*, K. B. Benson, ed., pp. 21.73-21.74, McGraw-Hill, New York, NY (1986).
3. B. D. Loughlin, "Monochrome and Color Visual Information Transmission," in *Television Engineering Handbook*, K. B. Benson, ed., pp. 4.4 - 4.5, McGraw-Hill (1986).
4. D. H. Pritchard, "Standards and Recommended Practices," in *Television Engineering Handbook*, K. B. Benson, ed., pp. 21.10-21.11, McGraw-Hill, New York, NY (1986).
5. S. C. Hsu, "The Kell Factor: Past and Present," *SMPTE Journal*, Vol. 95, pp. 206-214 (1986).
6. "EIA Standard EIA-343A, Electrical Performance Standards for High Resolution Monochrome Closed Circuit Television Camera," Electronic Industries Association, 2001 Eye Street, NW Washington, D.C. 20006 (1969).
7. See, for example, "Solving the Component Puzzle," Tektronix Book # 25W-7009-1, dated 9/90, Tektronix Inc., Beaverton, OR.
8. J. Hamalainen, "Video Recording Goes Digital," *IEEE Spectrum*, Vol. 32(4), pp. 76-79 (1995).
9. B. D. Loughlin, "Monochrome and Color Visual Information Transmission," in *Television Engineering Handbook*, K. B. Benson, ed., pp. 4.24-4.42, McGraw-Hill, New York, NY (1986).
10. L. J. Thorpe, "HDTV and Film - Issues of Video Signal Dynamic Range," *SMPTE Journal*, Vol. 100(10), pp. 780-795 (1991).
11. L. J. Thorpe, "HDTV and Film - Digitization and Extended Dynamic Range," *SMPTE Journal*, Vol. 102(6), pp. 486-497 (1993).
12. B. L. Lechner, "HDTV Status and Prospects," in *Seminar Lecture Notes*, 1995 SID International Symposium, pp. F4/4-F4/74, Society for Information Display, Santa Ana, CA (1995).

13. J. Blankevoort, H. Blom, P. Brouwer, P. Centen, B. vd Herik, R. Koppe, A. Moelands, J. v Rooy, F. Stok, and A. Theuwissen, "A High-Performance, Full-Bandwidth HDTV Camera Applying the First 2.2 Million Pixel Frame Transfer CCD Sensor," *SMPTE Journal*, Vol. 103(5), pp. 319-329 (1994).
14. K. B. Benson, ed., *Television Engineering Handbook*, Chapter 2 and pp. 22.29 - 22.39, McGraw-Hill, New York, NY (1986).
15. K. A. Parulski, L. J. D'Luna, and R. H. Hubbard, "A Digital Color CCD Imaging Systems using Custom VLSI Circuits," *IEEE Transactions on Consumer Electronics*, Vol. CE-35(8), pp. 382-388 (1989).
16. S. Nishikawa, H. Toyoda, V. Miyakawa, R. Asada, Y. Kitamura, M. Watanabe, T. Kiguchi, and M. Taniguchi, "Broadcast-Quality TV Camera with Digital Signal Processor," *SMPTE Journal*, Vol. 99(9), pp. 727-733 (1990).
17. Y. Takemura, K. Ooi, M. Kimura, K. Sanda, A. Kuboto, and M. Amano, "New Field Integration Frequency Interleaving Color Television Pickup System for Single Chip CCD Camera," *IEEE Transactions on Electron Devices*, Vol. ED-32(8), pp. 1402-1406 (1985).
18. H. Sugiura, K. Asakawa, and J. Fujino, "False Color Signal Reduction Method for Single Chip Color Video Cameras," *IEEE Transactions on Consumer Electronics*, Vol. 40(2), pp. 100-106 (1994).
19. J.-C. Wang, D.-S. Su, D.-J. Hwung, and J.-C. Lee, "A Single Chip CCD Signal Processor for Digital Still Cameras," *IEEE Transactions on Electron Devices*, Vol. 40(3), pp. 476-483 (1994).
20. D. H. Pritchard, "Standards and Recommended Practices," in *Television Engineering Handbook*, K. B. Benson, ed., pp. 21.46, McGraw-Hill, New York, NY (1986).
21. SMPTE 240M-1988 appears in *SMPTE Journal*, Vol. 98, pp. 723-725 September 1989 and can be obtained from SMPTE, 595 West Hartsdale Ave. White Plains, NY 10607.
22. A. N. Thiele, "An Improved law of Contrast Gradient for High Definition Television," *SMPTE Journal*, Vol. 103(1), pp. 18-25 (1994)
23. T. S. Lomheim and L. S. Kalman, "Analytical Modeling and Digital Simulation of Scanning Charge-Coupled Device Imaging Systems," in *Electro-Optical Displays*, M. A. Karim, ed., pp. 551-560, Marcel Dekker, New York (1992).
24. C. Sabolis "Seeing is Believing," *Photonics Spectra*, pp. 119-124, October 1993.
25. D. H. Sheingold, *Analog-Digital Conversion Handbook*, pp. 317-337, Prentice-Hall, Englewood Cliffs, NJ (1986).
26. See, for example, I. Csorba, *Image Tubes*, Howard Sams, Indianapolis, Indiana (1985).
27. Y. Talmi, "Intensified Array Detectors," in *Charge-Transfer Devices in Spectropscopy*, J. V. Sweedler, K. L. Ratzlaff, amd M. B. Denton, eds., Chapter 5, VCH Publishers, New York (1994).
28. I. Csorba, *Image Tubes*, pg. 209, Howard Sams, Indianapolis, Indiana (1985).
29. G. M. Williams, "A High-Performance LLLTV CCD Camera for Nighttime Pilotage," in *Electron Tubes and Image Intensifiers*, C. B. Johnson and B. N. Laprade, eds., SPIE Proceedings Vol. 1655, pp. 14-32 (1992).

6
CAMERA PERFORMANCE

Camera performance metrics are conceptually the same as array metrics (see Chapter 4, *Array Performance*, page 90). The camera is limited by the array noise floor and charge well capacity. The camera's fixed pattern noise (FPN) and photoresponse nonuniformity (PRNU) may be better than the array pattern noise when correction algorithms are present. Frame averaging can reduce the random noise floor and thereby appear to increase the camera's sensitivity.

Historically, performance metrics were created for broadcast systems. These metrics may be applied to scientific, machine vision, and military systems. Some of these metrics are not appropriate for specific applications and therefore may be considered as superfluous information.

Image intensifier and CCD array characteristics both affect the intensified CCD (ICCD) signal-to-noise ratio. ICCDs have a dynamic range that varies with the image intensifier gain. This permits fabrication of general purpose ICCDs that operate in lighting conditions that range from starlight to full sun light.

This chapter covers camera noise and the signal-to-noise ratio. It does not other issues such as frame rate requirements, blooming level, and color rendition. The symbols used in this book are summarized in the *Symbol List* (page xiv) and it appears after the *Table of Contents*.

6.1. STANDARD CAMERA

The standard camera contains a CCD array whose output has been formatted into standard video timing. The lens focuses reflected scene radiation onto the array.

6.1.1. SIGNAL

Referring to Figure 4-1, page 92, camera output voltage is

$$V_{CAMERA} = \frac{G_1 G q}{C} n_{pe} \quad (6\text{-}1)$$

The number of photoelectrons is

$$n_{pe} = \int_{\lambda_1}^{\lambda_2} R_q(\lambda) \frac{M_q(\lambda,T) A_D t_{INT}}{4 F^2 (1 + M_{OPTICS})^2} \tau_{OPTICS}(\lambda) \, d\lambda \quad (6\text{-}2)$$

For general applications, the object is considered to be far away so that M_{OPTICS} approaches zero. If the optical transmittance does not have any spectral features, then

$$n_{pe} = \frac{A_D \tau_{OPTICS} t_{INT}}{4 F^2} \int_{\lambda_1}^{\lambda_2} R_q(\lambda) M_q(\lambda,T) \, d\lambda \quad (6\text{-}3)$$

The photon level that saturates the charge wells is the camera's maximum input signal and $V_{CAMERA} = G_1 G q N_{WELL}/C$. Since the number of photoelectrons is inversely proportional to F^2, a f-number must be selected. Historically, film-based cameras are compared when the lens f-number is 5.6. Figure 6-1 illustrates the relative change in the maximum signal as the f-number is varied. For consumer video and industrial cameras, the integration is selected to match the video standard ($t_{INT} = 1/60$ for EIA 170). Automatic irises and variable integration times increase the maximum level.

Although the off-chip electronics is shown to have only one amplifier in Figure 4-1 (page 92), it may consist of several amplifiers. One of the amplifiers may have a variable gain that allows the user to increase the signal under low-light-level conditions. This gain is turned off (i.e., unity gain) when the maximum illumination is measured. Gain information is superfluous because the only requirement is to match a standard video output format. The manufacturer adds sufficient gain to achieve this value.

The minimum input signal suggests something about the lowest detectable signal. In a well-designed camera, the electronics does not add any significant noise and there is sufficient gain to measure the CCD array noise. The minimum signal is often specified with different optics than the optics specified for the maximum signal (Figure 6-2). Faster optics is a reasonable choice since these lenses are used in low-light-level conditions.

The camera manufacturer may amplify the signal so that the "minimum" signal provides full video (100 IRE units) or may use less gain so that the minimum signal provides, for example, 30% of the video signal (30 IRE units). As with the maximum signal, gain information is superfluous.

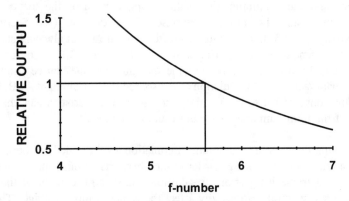

Figure 6-1. Relative maximum output normalized to f/5.6. The maximum input signal is often specified for a camera with an f/5.6 lens system.

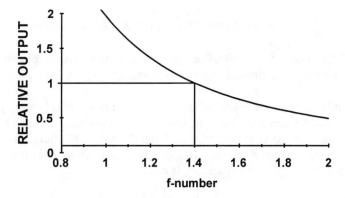

Figure 6-2. Relative minimum output normalized to f/1.4. A camera can appear to have very high sensitivity if a very low f-number lens is used.

158 CCD ARRAYS, CAMERAS, and DISPLAYS

Based on signal detection theory, the minimum illumination would imply that the signal-to-noise ratio is one. Because of its incredible temporal and spatial integration capability, the eye/brain system can perceive SNRs as low as 0.05. This is further discussed in Section 12.1.1., *Perceived Signal-to-Noise Ratio*. The definition of minimum illumination is author dependent. Its value may be (a) when the SNR is one, (b) when the video signal is, say, 30 IRE units, or (c) when an observer just perceives a test pattern. Therefore, comparing cameras based on "minimum" illumination capability should be approached with care.

Minimum and maximum illumination levels depend on the spectral output of the source and the spectral response of the detector (see Section 2.5. *Responsivity*, page 37). The source color temperature is not always listed but is a critical parameter for comparing systems. Although a CIE A source is used most often, the user should not assume that this was the source used by the camera manufacturer. Manipulation of the source (CIE illuminant A, B, C, D_{65}, or another unspecified source) and optical filter characteristics can change the values of the minimum and maximum illumination levels.

Dynamic range is the maximum illumination level divided by the minimum illumination level. It depends on lens f-number, iris setting, integration time, binning, and frame integration. In addition, the spectral content of the source and the array spectral responsivity affect the camera output voltage. Thus, the camera dynamic range can be quite variable depending on test conditions.

Example 6-1
MINIMUM PHOTOMETRIC SIGNAL ESTIMATION

A CCD array has a noise floor of 150 electrons rms. If placed into a camera, what is the estimated minimum detectable signal in lux?

The minimum signal occurs when the signal-to-noise ratio is one. That is, $n_{pe} = \langle n_{FLOOR} \rangle$. Several simplifying assumptions are required to obtain a back-of-the-envelope expression for the minimum signal. If the CIE illuminant D_{6500} is used, the spectral photon sterance is approximately constant over the spectral region, $\Delta\lambda$. If the quantum efficiency is constant,

$$M_q(\lambda) \approx \frac{\langle n_{FLOOR} \rangle}{\eta \, \Delta\lambda \, t_{INT}} \quad \frac{\text{photons}}{\text{sec} - \mu m} \qquad (6-4)$$

CAMERA PERFORMANCE 159

The energy of one photon is hc/λ. Then the spectral power at the detector is

$$M_p(\lambda) \approx \frac{\langle n_{FLOOR} \rangle}{\eta \, \Delta\lambda \, t_{INT}} \frac{hc}{\lambda} \quad \frac{watts}{\mu m} \qquad (6\text{-}5)$$

The illumination at the lens, whose area is A_O, is

$$\frac{\Phi_v}{A_O} = \frac{683}{A_O} \int_{0.38\,\mu m}^{0.75\,\mu m} V(\lambda) M_p(\lambda) \, d\lambda \quad lux \qquad (6\text{-}6)$$

Combining and adding the optical transmittance

$$\frac{\Phi_v}{A_O} \approx \frac{\langle n_{FLOOR} \rangle hc}{\eta \, \Delta\lambda \, \tau_{OPTICS} \, t_{INT}} \frac{683}{A_O} \int_{0.38}^{0.75} \frac{V(\lambda)}{\lambda} d\lambda \quad lux \qquad (6\text{-}7)$$

From Figure 4-3 (page 94) $\eta \approx 0.3$. For EIA 170 compatibility, $t_{INT} = 1/60$ sec, and the spectral bandwidth is $\Delta\lambda = (0.7 - 0.4)\,\mu m = 0.3\,\mu m$. Assume that the lens diameter is 1.5 cm ($A_O = 1.74 \times 10^{-4}\, m^2$). The photometric integral is equal to 0.2. The constant, hc, is approximately 2×10^{-19} joule-μm. The optical transmittance is approximately 0.9 and the minimum signal, $\langle n_{FLOOR} \rangle$, is 150 electrons. This provides an approximate theoretical minimum detectable signal of 0.017 lux.

Example 6-2
MAXIMUM SIGNAL ESTIMATION

The camera described in Example 6-1 has a well capacity of 100,000 electrons. What is the estimated maximum input signal?

The dynamic range is $N_{WELL}/\langle n_{FLOOR} \rangle$ or 56.5 dB. The maximum input value is estimated with an f/5.6 lens whereas the minimum level is estimated with an f/1.4 lens. The higher f-number is represented by a smaller diameter lens (smaller area). Then the saturation equivalent exposure is

$$SEE = \frac{100,000}{150} \left(\frac{5.6}{1.4}\right)^2 0.017 \approx 181 \quad lux \qquad (6\text{-}8)$$

Higher values are possible by using a shorter integration time. An automatic iris effectively increases the f-number and thereby increases the saturation equivalent exposure.

6.1.2. CAMERA SIGNAL-TO-NOISE RATIO

For large targets, the camera SNR is similar to the array SNR (see Section 4.3., *Array Signal-to-Noise Ratio*, page 119). Noise is inserted by the detector and by the amplifiers (see Figure 4-7, page 105). The electronic band pass, $H_{SYS}(f_{elec})$, which is determined by the electronics MTF, modifies the noise spectra. The MTFs are described in Chapter 10, *System MTF*.

When measured at the camera output, the SNR for large targets is

$$SNR = \frac{\dfrac{G_1 G q}{C} \dfrac{A_D}{4F^2} \int_{\lambda_1}^{\lambda_2} R_q(\lambda) M_q(\lambda,T) t_{INT} \tau_{OPTICS}(\lambda) d\lambda}{\sqrt{\int_0^\infty S(f_{elec}) |H_{SYS}(f_{elec})|^2 df_{elec}}} \quad (6\text{-}9)$$

$S(f_{elec})$ is the noise power spectral density and includes all the noise sources: shot noise, noise floor, reset noise, pattern noise, quantization noise, amplifier white noise and amplifier 1/f noise. The subscript *elec* emphasizes that electrical frequency applies to the amplifiers and subsequent electronic filters.

Equation 6-9 is the generalized SNR equation. If the amplifier noise is negligible, then it reduces to the familiar form:

$$SNR = \frac{n_{pe}}{\langle n_{SYS} \rangle} = \frac{\dfrac{A_D}{4F^2} \int_{\lambda_1}^{\lambda_2} R_q(\lambda) M_q(\lambda,T) t_{INT} \tau_{OPTICS}(\lambda) d\lambda}{\langle n_{SYS} \rangle} \quad (6\text{-}10)$$

For a specific target spatial frequency, f_o, the generalized SNR is

$$SNR = \frac{\dfrac{G_1 G q}{C} \dfrac{A_D}{4F^2} \int_{\lambda_1}^{\lambda_2} R_q(\lambda) M_q(\lambda,T) MTF_{SYS}(f_o) t_{INT} \tau_{OPTICS}(\lambda) d\lambda}{\sqrt{\int_0^\infty S(f_{elec}) |H_{SYS}(f_{elec})|^2 df_{elec}}} \quad (6\text{-}11)$$

CAMERA PERFORMANCE 161

Where MTF_{SYS} affects the (periodic) signal intensity. MTF_{SYS} is the total system MTF including the optics, detector, and electronics. H_{SYS} is the electronics MTF only. Again, if the amplifiers do not add any appreciable noise, the equation reduces to the familiar form:

$$SNR(f_o) \approx \frac{n_{pe} MTF_{sys}(f_o)}{\langle n_{SYS} \rangle} \qquad (6\text{-}12)$$

The application of Equation 6-12 is further discussed in Chapter 12, *Minimum Resolvable Contrast*. Point source detection calculations are more difficult and the reader is referred to the literature[1,2].

6.1.3. NOISE EQUIVALENT INPUT

In Example 6-1, page 158, the minimum signal was specified in photometric units. In the more general radiometric case, the input that provides SNR = 1 is the noise equivalent input (NEI). Applying the same approximations to Equation 6-10 provides the NEI in photon flux:

$$NEI \approx \frac{4F^2}{A_D t_{INT} \eta \Delta\lambda \tau_{optics}} \langle n_{SYS} \rangle \quad \frac{photons}{sec\text{-}cm^2\text{-}\mu m} \qquad (6\text{-}13)$$

When the SNR is unity, the number of photons is small and photon noise is negligible (see Figure 4-12b, page 117). If the detector is cooled, dark current is minimized. With PRNU and FPN correction, the minimum NEI is

$$NEI \approx \frac{4F^2}{A_D t_{INT} \eta \Delta\lambda \tau_{OPTICS}} \langle n_{FLOOR} \rangle \quad \frac{photons}{sec\text{-}cm^2\text{-}\mu m} \qquad (6\text{-}14)$$

If TDI is used, then N_{TDI} is multiplied in the denominator. With binning or super pixeling, the number of pixels combined is multiplied in the denominator.

6.1.4. NOISE EQUIVALENT REFLECTANCE

The NEI assumes that the target is against a black background. For real scenes, both the target and background create signals. The target is only visible when a sufficient reflectance difference exists. For detecting objects at long ranges, the atmospheric transmittance loss must be included. The signal difference between the target and its background is

$$\Delta n_{pe} = \int_{\lambda_1}^{\lambda_2} R_q(\lambda) \frac{(\rho_T - \rho_B) M_q(\lambda) t_{INT} A_D}{4F^2} \tau_{OPTICS}(\lambda) T_{ATM}(\lambda) d\lambda \quad (6\text{-}15)$$

Both the target and background are illuminated by the same source. The signals from both depend on the spectral reflectances that are taken as constants over the spectral region of interest. Using the same simplifying assumptions as before, solving for $\rho_T - \rho_B$ and calling this value the noise equivalent reflectance difference (NE$\Delta\rho$) provides

$$NE\,\Delta\rho \approx \frac{4F^2}{A_d t_{INT}\, \eta\, M_q(\lambda_o)\, \Delta\lambda\, \tau_{OPTICS}\, T_{ATM}} \langle n_{SYS} \rangle \quad (6\text{-}16)$$

T_{ATM} is the total atmospheric transmittance and depends on range. T_{ATM} is usually expressed as $T_{ATM} = \exp(-\sigma_{ATM} R)$. Then

$$NE\,\Delta\rho \approx \frac{4F^2 e^{\sigma_{ATM} R}}{A_D t_{INT}\, \eta\, M_q(\lambda_o)\, \Delta\lambda\, \tau_{OPTICS}} \langle n_{SYS} \rangle = \frac{NEI}{M_q(\lambda_o)} e^{\sigma_{ATM} R} \quad (6\text{-}17)$$

Thus, NE$\Delta\rho$ depends on the source illumination, system noise, and the atmospheric conditions. As the range increases, the target reflectance must increase to keep the SNR at one. The maximum range occurs when the NE$\Delta\rho$ is one. Then

$$NEI = M_q(\lambda_o)\, e^{-\sigma_{ATM} R} \quad (6\text{-}18)$$

$M_q(\lambda)[\rho_T(\lambda) - \rho_B(\lambda)]$ is a complicated function that depends on the cloud cover, atmospheric conditions, sun elevation angle, surface properties of the target, and the angle of the target with respect to the line-of-sight and sun angle.

These back-of-the envelope approximations are still not complete. They assume that the total received energy is from the target only. In fact, over long path lengths, path radiance contributes to the energy detected. Path radiance, specified by visibility, reduces contrast. The additional light reflected by the atmospheric particulates introduce photon noise that is independent of the signal from the target. It is quantified by the sky-to-background ratio (discussed in Section 12.6.1, *Contrast Transmittance*).

6.2. INTENSIFIED CCD CAMERA

The construction of the ICCD was described in Section 5.4.2., *Intensified CCDs*, page 149. As illustrated in Figure 6-3, the photocathode creates photoelectrons that are amplified by the MCP. The amplified electrons are converted back into photons by the phosphor screen. These photons are relayed to the CCD by either a fiber-optic bundle or a relay lens. The CCD creates the photoelectrons that are measured. Each conversion has a quantum efficiency less than unity and each optical element yields a finite transmittance loss.

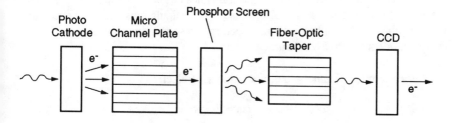

Figure 6-3. Multiple photon-electron conversions occur in an ICCD.

While the image intensifier amplifies the image, the SNR may be reduced due to quantum losses and additive noise. For moderate light levels, the ICCD offers no advantage in SNR performance over the standard CCD camera. Its key attribute is the ability to temporally gate the scene.

6.2.1. SIGNAL

The ICCD uses a lens to image the target onto the intensifier face plate. The photon flux reaching the photocathode is

$$\Phi_{CATHODE} = \frac{M_q(\lambda,T) A_{CATHODE}}{4 F^2 (1 + M_{OPTICS})^2} \tau_{OPTICS} \tau_{WINDOW} \qquad (6\text{-}19)$$

Using the same approximations as given in the preceding section, the number of signal electrons generated by the photocathode is

$$n_{CATHODE} = \frac{\eta_{CATHODE} M_q(\lambda_o) \Delta\lambda}{4 F^2} t_{INT} \tau_{OPTICS} A_{PIXEL} \qquad (6\text{-}20)$$

A_{PIXEL} is the projected area of a CCD pixel onto the photocathode. Geometrically, it is the CCD detector area multiplied by the magnification of the fiber-optic bundle. However, A_{PIXEL} depends on fiber-optic bundle characteristics and microchannel pore size. It may be somewhere between 50% of the CCD detector geometric projected area to several times larger[3]. As such, it represents the largest uncertainty in signal-to-noise ratio calculations. λ_o is the center wavelength of the photocathode spectral responsivity and $\Delta\lambda$ is the photocathode's spectral bandwidth.

After the microchannel plate, the number of electrons is

$$n_{MCP} = G_{MCP} \, n_{CATHODE} \qquad (6\text{-}21)$$

These electrons are then converted to photons by the phosphor screen

$$n_{SCREEN} = \eta_{SCREEN} n_{MCP} \qquad (6\text{-}22)$$

The photons travel to the CCD via a fiber-optic bundle whose effective transmittance is T_{fo}. This is a composite value that includes the fiber-optic transmittance and a coupling loss.

$$n_{PHOTON\text{-}CCD} = T_{fo} n_{SCREEN} \qquad (6\text{-}23)$$

The number of electrons sensed in the CCD is proportional to its quantum efficiency

$$n_{pe} = \eta_{CCD} n_{PHOTON-CCD} \quad (6\text{-}24)$$

or

$$n_{CCD} = \eta_{CCD} \eta_{SCREEN} \eta_{CATHODE} T_{fo} G_{MCP} \left[\frac{M_q(\lambda_o) \Delta\lambda\, t_{INT}\, A_{PIXEL}}{4F^2} \right] \quad (6\text{-}25)$$

For a lens coupled system, T_{FO} is replaced by an effective relay lens transmittance[4]. Each has its own merits. Transmission losses in a fiber-optic bundle increase proportionally to the square of minification ratio. This leads to a practical minification of about 2:1. Light is lost because the screen is a Lambertian source whereas the fiber-optic can only accept light over a narrow cone (acceptance angle). On the other hand, lenses are large and bulky and may provide a throughput as low as 2%. Relay lenses also suffer from vignetting. Here, the "relayed" intensity decreases as the image moves from the center to the periphery of the image intensifier tube.

Since the lens coupling is less efficient than the fiber-optic coupling, the number of electrons produced is lower with the relay lens. The fiber-optic coupled system, with its higher effective transmittance and better sensitivity, can detect lower signals.

6.2.2. ICCD NOISE

The photocathode produces both photon noise and dark noise.

$$\langle n^2_{PC-SHOT} \rangle = \langle n^2_{CATHODE} \rangle + \langle n^2_{PC-DARK} \rangle \quad (6\text{-}26)$$

The MCP amplifies these signals and adds excess noise so that

$$\langle n^2_{MCP} \rangle = k_{MCP} G^2_{MCP} \langle n^2_{PC-SHOT} \rangle \quad (6\text{-}27)$$

k_{MCP} is typically taken as 1.8 but varies with bias voltage[5,6]. k_{MCP} is also reported as an SNR degradation.

166 CCD ARRAYS, CAMERAS, and DISPLAYS

The noise power is transformed into photons by the phosphor screen

$$\langle n^2_{SCREEN}\rangle = \eta^2_{SCREEN}\langle n^2_{MCP}\rangle \qquad (6\text{-}28)$$

It is attenuated by the fiber-optic bundle

$$\langle n^2_{CCD-PHOTON}\rangle = T^2_{fo}\langle n^2_{SCREEN}\rangle \qquad (6\text{-}29)$$

This value is converted to electrons by the CCD and the CCD noise is added

$$\langle n^2_{CCD}\rangle = \eta^2_{CCD}\langle n^2_{CCD-PHOTON}\rangle + \langle n^2_{CCD-DARK}\rangle + \langle n^2_{FLOOR}\rangle + \langle n^2_{PRNU}\rangle \qquad (6\text{-}30)$$

The total noise is

$$\langle n^2_{CCD}\rangle = \eta^2_{CCD}T^2_{fo}\eta^2_{SCREEN}kG^2_{MCP}\langle n^2_{PC-SHOT}\rangle + \langle n^2_{CCD-SHOT}\rangle + \langle n^2_{FLOOR}\rangle + \langle n^2_{PRNU}\rangle$$
$$(6\text{-}31)$$

6.2.3. ICCD SNR

The resultant signal-to-noise ratio is

$$SNR = \frac{\eta_{CCD}\eta_{SCREEN}T_{fo}G_{MCP}n_{CATHODE}}{\sqrt{\eta^2_{CCD}T^2_{fo}\eta^2_{SCREEN}kG^2_{MCP}\langle n^2_{PC-SHOT}\rangle + \langle n^2_{CCD-SHOT}\rangle + \langle n^2_{FLOOR}\rangle + \langle n^2_{FPN}\rangle}} \qquad (6\text{-}32)$$

Several limiting cases are of interest. When the gain is very high, the intensified camera noise overwhelms the CCD noise:

$$SNR = \frac{n_{CATHODE}}{\sqrt{k\left[\langle n^2_{CATHODE}\rangle + \langle n^2_{PC-DARK}\rangle\right]}} \qquad (6\text{-}33)$$

The SNR is independent of gain and CCD characteristics. Under high gain conditions, an expensive, high quantum efficiency or thinned back sided illuminated CCD is not necessary. Cooling the photocathode reduces the dark current:

$$SNR = \sqrt{\frac{n_{CATHODE}}{k}} = \sqrt{\frac{\eta_{CATHODE} M_q(\lambda_o) \Delta \lambda}{4 F^2 k} t_{INT} \tau_{OPTICS} A_{PIXEL}} \quad (6\text{-}34)$$

High gain is used for photon counting applications. SNR, in this case, is limited by the photocathode quantum efficiency, the excess noise factor, and the optical system.

Under moderate gain situations where the CCD noise is greater than the intensifier noise,

$$SNR = \frac{\eta_{CCD} \eta_{SCREEN} T_{fo} G_{MCP} n_{CATHODE}}{\sqrt{\langle n^2_{CCD\text{-}SHOT} \rangle + \langle n^2_{FLOOR} \rangle + \langle n^2_{PRNU} \rangle}} \quad (6\text{-}35)$$

Comparing this with Equation 6-12 (page 161), the ICCD will have lower SNR than a comparable standard camera. The advantage of the ICCD under these conditions is the ability to temporally gate the incoming photon flux.

Example 6-3
IMAGE INTENSIFIER NOISE

Is there a relationship between the image intensifier noise equivalent input and the ICCD noise?

The noise in an image intensifier is due to random emissions (dark current noise) at the photocathode. When referred to the input (SNR = 1), it is called the equivalent background illumination (EBI). Using the same methodology as in Example 6-1, the EBI for an area A_{PIXEL} is

$$EBI = M_q(\lambda_o) \Delta \lambda t_{INT} A_{PIXEL} = \frac{4\sqrt{k} F^2 \langle n_{PC\text{-}DARK} \rangle}{\eta_{CATHODE} \tau_{OPTICS}} \text{ photons} \quad (6\text{-}36)$$

168 CCD ARRAYS, CAMERAS, and DISPLAYS

The ICCD will have the same equivalent background input under high gain situations. Cooling the image intensifier reduces the EBI and it may be negligible in gated applications. Under low gain situations, the CCD noise dominates.

Example 6-4
CCD versus ICCD

Under moderate gain situations, what is the relative SNR of the ICCD to the CCD?

The ratio of SNRs is

$$\frac{SNR_{ICCD}}{SNR_{CCD}} = \eta_{SCREEN} T_{fo} G_{MCP} \eta_{photocathode} \qquad (6\text{-}37)$$

The P20 phosphor screen has a peak emission at 0.56 μm with a conversion efficiency of 0.063. An S20 bialkali photocathode's efficiency is 0.2 at 0.53 μm. T_{fo} is about 0.25. Combined, these factors give 0.00315. The MCP gain must be greater than 1/0.00315 = 317 for the SNR_{ICCD} to be greater than SNR_{CCD}.

6.3. REFERENCES

1. G. C. Holst, *Electro-Optical Imaging System Performance*, pp. 48-54, JCD Publishing (1995).
2. M. A. Sartor, "Characterization and Modeling of Microchannel Plate Intensified CCD SNR Variations with Image Size," in *Electron Tubes and Image Intensifiers*, C. B. Johnson and B. N. Laprade, eds., SPIE Proceedings Vol. 1655, pp. 74-84 (1992).
3. Mark Sartor, Xybion Corporation, private communication.
4. Y. Talmi, "Intensified Array Detectors," in *Charge-Transfer Devices in Spectroscopy*, J. V. Sweedler, K. L. Ratzlaff, and M. B. Denton, eds., Chapter 5, VCH Publishers, New York (1994).
5. I. Csorba, *Image Tubes*, pp. 120-124, Howard Sams, Indianapolis, Indiana (1985).
6. R. J. Hertel, "Signal and Noise Properties of Proximity Focused Image Tubes," in *Ultrahigh Speed and High Speed Photography, Photonics, and Videography*, G. L. Stradling, ed., SPIE Proceedings Vol. 1155, pp. 332-343 (1989).

7
CRT-BASED DISPLAYS

Displays are used for consumer television receivers, computer monitors, scientific applications, and military applications. A display may be called a monitor, visual display unit (VDU), video display terminal (VDT) or, simply a television receiver.

The major performance requirements are listed in Table 7-1. This table is not meant to be all inclusive. It just indicates that different users have different requirements. As displays are built for each user, they are specified in units familiar to that user. For example, consumer displays will not have the MTF or spot size listed. These parameters are provided for militarized displays.

Table 7-1
DISPLAY REQUIREMENTS

APPLICATION	REQUIREMENTS
Consumer TV	"Good" image quality
Computer	Alphanumeric character legibility "Good" graphical capability
Scientific and military	High MTF Fixed gamma

Television receivers have been available for over 50 years and the technology has matured. They are considered adequate and little improvement in cathode ray tube (CRT) technology is expected. Advances in television displays currently focus on high definition television, flat panel displays, the legibility of alphanumeric characters, and computer graphics capability. As a result, display specifications are a mix of CRT terminology, video transmission standards, alphanumeric character legibility, and graphics terminology.

Addressability is a characteristic of the display controller and represents the ability to select and activate a unique area on the screen. It defines how precisely an electron beam can be positioned on the CRT screen. Addressability is given as the number of discrete lines per picture height or pixels per unit

170 CCD ARRAYS, CAMERAS, and DISPLAYS

length. With this definition, the inverse of addressability is the center-to-center spacing between adjacent pixels. The number of pixels that can exist depends on the spot diameter and the electronic bandwidth.

Resolution suggests something about the highest spatial frequency that can be seen. As an electron beam move across the screen, its amplitude is modulated. The maximum rate at which the beam can be modulated from full off to full on defines the maximum number of (monitor) pixels that can be written in a row. The beam moves down vertically to the next line and again travels left to right. The distance the beam moves down is the vertical extent of a pixel. While this can be any distance, it is usually selected so that the individual raster lines cannot be seen. Resolution and addressability specify monitor performance. They are combined into the resolution/addressability ratio (RAR). The RAR provides guidance to monitor design.

System resolution depends on the camera response and limitations imposed by the video standard, display characteristics, and observer acuity. High resolution displays do not offer any "extra" resolution from a system point of view if display is not the limiting factor. Rather, high-resolution displays just insure that all the information available is displayed. In many applications, *system* resolution is limited by the human eye. If the observer is too far from the screen, not all of the image detail can be discerned.

In almost all cases, the output device sets the system dynamic range. Often, the display is the limiting factor in terms of image quality and resolution. No matter how good the camera is, if the display resolution is poor, then the overall system resolution is poor.

Monitor pixels are not related to the camera pixels. Each is generated by the respective designs. CCD cameras with a standard video output may not have the resolution suggested by the video standard. For example, a camera containing an array of 320 x 240 pixels may have its output formatted into EIA 170 timing. This standard suggests that it can support approximately 437 x 485 pixels (see Table 5-2, page 129). The signal may be digitized by a frame grabber that creates 1000 x 485 pixels. The image size is still 320 x 240 pixels.

Flat panel displays are an emerging technology with liquid crystal display (LCD) prevalent in notebook (laptop) computers. LCDs require only a fraction of the power required by cathode ray tubes and therefore lend themselves to battery operated systems. However, CRTs will probably dominate the display market due to low cost, high resolution, wide color gamut, high luminance, and long life.

The commercially important flat panel displays are the light emitting diodes (LEDs) display, AC plasma display panel (AC PDP), DC plasma display panel (DC PDP), AC thin film electroluminescent (ACTFEL) display, DC thin film electroluminescent (DCTFEL) display, vacuum fluorescent display (VFD), and the liquid crystal display (LCD).

LCD arrays can either be active matrix (AMLCD) or passive arrays. The active matrix contains switching devices on each pixel to maintain high brightness and high contrast control. The liquid crystals may be either twisted-nematic (TN) or supertwisted-nematic (STN). The details of these displays can be found in references 1 through 5. Although this chapter discusses CRTs, many concepts apply to all displays.

Emerging technology, new designs, and characterization methods can be found in the *Society of Motion Picture and Television Engineers Journal* and the *Proceedings of the Society of Information Display* symposia. The Society of Information Display offers numerous tutorials, complete with published notes, at their symposia.

Following the industry convention, display sizes and viewing distances are expressed in inches. The remainder of the book uses the metric system. The symbols used in this book are summarized in the *Symbol List* (page xiv) and it appears after the *Table of Contents*.

7.1. THE OBSERVER

The observer is the starting point for display design. Design is based on both perceptual and physical attributes (Table 7-1). For most applications, the design is driven by perceptual parameters. These parameters are related, somewhat, to the physical parameters. For example, sharpness is related to MTF. But the precise relationship has not been quantified. While the physical parameters are important to all applications, they tend to be provided only for scientific and military displays. Table 7-1 and Table 7-2 illustrate different ways of specifying the same requirement: the need for a high-quality display.

Maximum image detail is only perceived when the observer is at an optimum viewing distance. If too far away, detail beyond the eye's limiting resolution. As he moves to the display, the image does not become clearer because there is no further detail to see. Display size and resolution requirements are based on an assumed viewing distance.

Table 7-2
PERCEPTIBLE AND PHYSICAL PARAMETERS

PERCEPTUAL PARAMETERS	PHYSICAL PARAMETERS
Brightness Contrast Sharpness Color rendition Flicker	Luminance Resolution Uniformity Addressability Gamma Color saturation Color accuracy Color convergence (CRT) Distortion (CRT) Refresh rate MTF

For monochrome systems, the optimum viewing distance occurs when the raster pattern is barely perceptibility. At closer distances, the raster pattern becomes visible and interferes with image interpretation. That is, it is considered annoying by most. For color CRTs, the observer must be far enough away so that the individual dots of the three primaries are imperceptible. Otherwise, the individual color spots become visible and the impression of full color is lost.

For design purposes, perceptibility occurs when targets are larger than 1 arc minute. This corresponds to the normal visual acuity of 20/20. This design criterion is quite good. An observer can adjust his distance accordingly. An observer with poorer acuity (e.g., 20/40) will simply move closer to the screen whereas someone with better acuity will probably move further away. Most systems are observer-limited in that the observer cannot see all the detail because he is typically too far from the display.

Figure 7-1 illustrates the optimum viewing distance as a function of display diagonal size for monochrome displays. Television receivers and many computer monitors are specified by the diagonal dimension. Each active line (raster line) subtends 1 arc minute (0.291 mrad). At closer distances, the individual raster lines may become visible. The viewing ratio is the viewing distance divided by the display height. The typical viewing ratio for computer monitors is three.

Television viewers sit about seven times (viewing ratio is seven) the picture height or 4.2 times the diagonal picture dimension. Knowing the dimensions of a typical living room (sofa on one side, television on the other) suggests that consumer televisions should have a diagonal picture size of 21 to 25 inches. These were the most popular television receiver picture sizes. Sizes have changed reflecting changing in television viewing habits.

A major design consideration is the elimination of flicker. For progressive scan imagery, the frame rate (refresh rate) must be near 70 Hz to avoid flicker. Flicker and other design features are not covered here but can be found in numerous books[2-5].

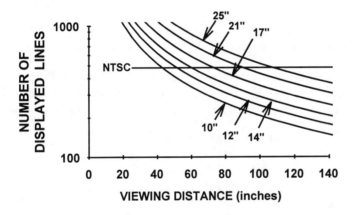

Figure 7-1. Viewing distance as a function of CRT diagonal picture size and number of displayed lines per picture height for monochrome displays. Most CRTs have a 4:3 aspect ratio so that the vertical extent is 60% of the diagonal.

7.2. CRT OVERVIEW

Figure 7-2 illustrates a typical color cathode ray tube display. Although a color display is illustrated, the block diagram applies to all CRT displays. The input video signal is decoded into its components and then digitized. While digital electronics may be used to process the signal, the CRT is an analog device. Similarly, displays accepting digital inputs must convert the signal into an analog voltage to drive the CRT.

174 CCD ARRAYS, CAMERAS, and DISPLAYS

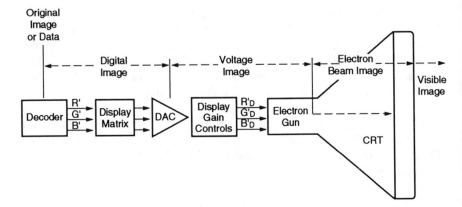

Figure 7-2. Typical display. Although a color display is illustrated, the block diagram applies to all CRT displays.

The display matrix provides the mapping between the standard video color components and the actual colors emitted by the phosphors on the screen. The color correction circuitry is an algorithm that can be represented as a matrix

$$\begin{pmatrix} R_2 \\ G_2 \\ B_2 \end{pmatrix} = \begin{pmatrix} k_{11} & k_{12} & k_{13} \\ k_{21} & k_{22} & k_{23} \\ k_{31} & k_{32} & k_{33} \end{pmatrix} \begin{pmatrix} R_1 \\ G_1 \\ B_1 \end{pmatrix} \qquad (7\text{-}1)$$

The digital-to-analog converter (DAC) provides the analog voltages that modulate the electron beam.

The electron beams paint an image on the CRT phosphor in the scan format shown in Figure 5-2, page 127. The color of the visible image depends on the specific phosphors used. Electron beam density variations, which represent scene intensity variations, create variations in the visible image intensity.

Monochrome displays have a single electron gun and a single phosphor. A color CRT must produce three images (three primary colors) to give a full range of color. To do so requires three electron guns and three phosphors. The three electron beams scan the CRT screen in the same way as a monochrome beam, but arrive at the screen at slightly different angles. The beams strike three different phosphors dots. When the dots (*sub*pixels) are beyond the eye's limiting resolution, the eye blends them to create the illusion of a full color spectrum.

CRT-BASED DISPLAYS 175

The output brightness of a CRT depends on the grid voltage and follows a power law relationship. The slope of this curve, on a log-log scale, is the display gamma (γ).

$$L_{DISPLAY} = k(V_{GRID})^\gamma \qquad (7\text{-}2)$$

Gamma can range[6] from one to five. For design purposes, NTSC standardized gamma at 2.2. PAL and SECAM standardized gamma at 2.8.

To provide a linear relationship in the complete camera-display system, the inverse gamma is inserted in the consumer camera as an image processing algorithm (see Section 5.3.3., *Gamma Compensation*, page 142). For scientific applications and computer monitors, the displays will have an internal gamma correction circuit. Here, the user simply supplies the signal with no gamma compensation (e.g., 1/γ = 1). For color displays, each color will have its own gamma look-up-table. This operation is performed in the display matrix.

Digital displays provide excellent alphanumeric and graphic imagery. There is no distortion and the pixels are accurately located because there is no jitter in timing signals compared to what might be expected with EIA 170. "Digital" seems to imply high resolution but the displayed resolution is limited by the electron beam spot size. Even though called "digital," the CRT is an analog device.

7.2.1. MONOCHROME DISPLAYS

In many industrial and scientific applications, monochrome cameras are used. The military uses monochrome monitors extensively to display imagery from non-visible scenes (e.g., thermal and radar imagery). Monochrome displays are found in cockpits and medical imaging instruments. For television receivers, these are simply called black-and-white TVs. Depending on the phosphor used, monochrome displays can provide nearly any color desired. Certain colors (e.g., orange) tends to reduce eye strain. Others (e.g., blue) tend not to disrupt scoptic vision adaption. Although computers have color monitors, word processing is often performed in a monochrome format.

Black and white scenes can be imaged on a color monitor by assigning a color palette (false color or pseudo color) to various intensity levels. This does not increase resolution but draws the observer's attention to selected features. The eye can easily differentiate color differences but may have trouble distinguishing small intensity differences.

7.2.2. COLOR DISPLAYS

A color monitor has three electron beams that scan the CRT face plate. Near the face plate is a metal screen or shadow mask that has a regular pattern of holes (Figure 7-3). The three guns are focussed onto the same hole and the emerging beams strike the CRT face plate at different locations. At each location there is a different phosphor. One fluoresces in the blue, one in the red, and one in the green region of the spectrum. The arrangement is such that the beam corresponding to the desired color will strike only the phosphor dot producing that color. The electron guns emit only electrons and have no color associated with them. Color is only created when the beam hits the phosphor even though the guns are called red, green, and blue guns.

Figure 7-4 illustrates one possible arrangement of color dots. The dot pitch is the center-to-center spacing between adjacent like-colored dots. The three dots create a triad. Since each electron beam passes through the same hole, the mask pitch is the same as the triad pitch. Table 7-3 relates the triad dot pitch to the more common definitions of CRT resolution.

Table 7-3
RESOLUTION OF COLOR DISPLAYS

RESOLUTION	TRIAD DOT PITCH
Ultra-high	< 0.27 mm
High	0.27 - 0.32 mm
Medium	0.32 - 0.48 mm
Low	> 0.48 mm

By varying the ratio of red, green, and blue, a variety of effective color temperatures can be achieved. The video signal is simply a voltage with no color temperature associated with it. The display electronics can be adjusted so that the display brightness appears to be somewhere between 3200 K and 10,000 K. Most displays are preset to either 6500 K or 9300 K. As the color temperature increases, whites appear to change from a yellow tinge to a blue tinge. The perceived color depends on the adapting illumination (e.g., room lights). Setting the color temperature to 9300 K provides aesthetically pleasing imagery and this setting is unrelated to the actual scene colors (see Section 5.3.2., *Color Correction*, page 138).

CRT-BASED DISPLAYS 177

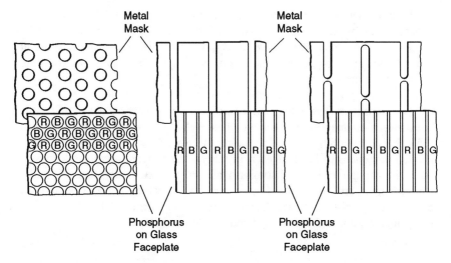

Figure 7-3. Color CRT mask patterns. (a) Dotted, (b) stripes, and (c) slotted. Most consumer television receivers use slots and computer displays use dots. The shadow mask design varies with manufacturer.

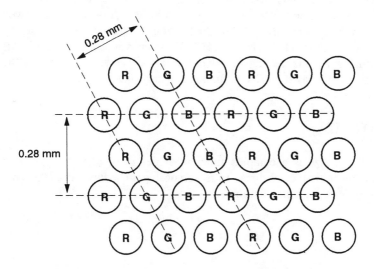

Figure 7-4. Arrangement of phosphor dots on a CRT. There is one shadow mask hole for each triad. In 1995, most computer monitors had a triad pitch of 0.28 mm.

178 CCD ARRAYS, CAMERAS, and DISPLAYS

Usually there is no relationship between the color temperature and contrast of a scene and what is seen on a display. This is not the fault of the camera and display manufacturers; they strive for compatibility. The observer, who usually has no knowledge of the original scene, will adjust the display for maximum visibility and aesthetics.

For consumer television and consumer computer graphics applications, adequate color imagery is obtained when each primary has 8 bits of intensity (256 shades of gray). The combination creates 24 bits or 16 million colors. Here, each intensity level of each primary is considered a color. It does not indicate that the eye can distinguish all 16 million - only that 16 million combinations are possible. Finer intensity increments may be used in commercial graphics displays. These systems may operate at 10 bits or 12 bits per primary color and provide 2^{30} and 2^{36} "colors" respectively.

7.2.3. HDTV

As the name implies, high definition television (HDTV) provides more resolution than conventional television receivers. The HDTV display has a 16:9 aspect ratio compared to the NTSC 4:3 aspect ratio. It is designed to be viewed at three times the picture height. While the horizontal visual angle subtended by NTSC systems is about 11 degrees, the HDTV subtends about 33 degrees. This larger format gives the feeling of teleprescence - that you are there! The operation of a CRT-based HDTV is the same as any other CRT.

7.3. SPOT SIZE

Display spot size is a complex function of design parameters, phosphor choice, and operating conditions. It is reasonable to assume that the spot intensity profile is Gaussian distributed:

$$L(r) = A e^{-\frac{1}{2}\left(\frac{r}{\sigma_{SPOT}}\right)^2} \tag{7-3}$$

The spot is radially symmetric with radius r.

As illustrated in Figure 7-5, the spot diameter is defined as the full width at half-maximum (FWHM) and is:

$$S = FWHM = \sqrt{8\ln(2)}\ \sigma_{SPOT} = 2.35\ \sigma_{SPOT} \tag{7-4}$$

Then

$$L(r) = A e^{-4\ln(2)\left(\frac{r}{S}\right)^2} \tag{7-5}$$

The advantage of assuming a Gaussian spot is that many resolution measures are easily compared and the display modulation transfer function is uniquely specified (discussed in Section 10.6., *Display*).

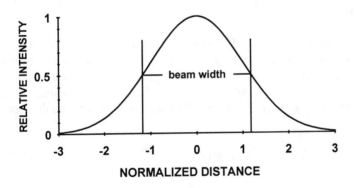

Figure 7-5. The Spot diameter is the full beam width at one-half maximum intensity. The profile has been normalized to σ_{SPOT}.

While the shrinking raster method is listed as a resolution test method, its primary value is in determining the FWHM spot size. With the shrinking raster method, every line is activated (Figure 7-6). If the lines are far apart, the valleys between the lines are visible and the raster pattern is obvious. The lines are slowly brought together until a flat field is encountered. That is, the entire screen appears to have a uniform intensity. The resolution is the number of displayed lines divided by the picture height of the shrunk raster.

180 CCD ARRAYS, CAMERAS, and DISPLAYS

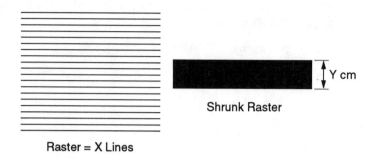

Figure 7-6. Shrinking raster resolution test method. The raster is shrunk until the discrete lines can no longer be discerned. The resolution is R^R = X/Y lines/cm.

As the lines coalesce, the MTF drops. Under nominal viewing conditions, experienced observers can no longer perceive the raster when the luminance variation (ripple) is less than 5%. The line spacing has been standardized[8] to $2\sigma_{SPOT}$ (Figure 7-7). Given a shrunk raster resolution of R_R lines/cm, the line spacing in the shrunk raster is $1/R_R$ cm. Then $\sigma_{SPOT} = 1/2R_R$ cm and the spot size is:

$$S = FWHM = \frac{2.35}{2R} \qquad (7\text{-}6)$$

While a flat field condition is desirable, the line spacing may be larger. The shrinking raster methodology is applicable for CRTs only and cannot be used for flat panel displays since the "raster" is fixed in those displays.

While the spot profile tends to be flat-topped in many displays, a Gaussian approximation is adequate if the standard deviation, σ_{SPOT}, is calculated from the 5% intensity level[9]:

$$\sigma_{SPOT} = \frac{FW @ 5\% \text{ level}}{\sqrt{8\ln(20)}} \approx \frac{FW @ 5\% \text{ level}}{4.9} \qquad (7\text{-}7)$$

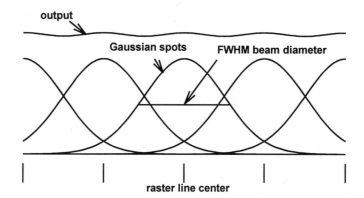

Figure 7-7. When the line spacing is $2\sigma_{SPOT}$ the spots appear to merge and this produces a flat field. The peak-to-peak ripple is 3% and the MTF associated with the residual raster pattern is 0.014. The output is the sum of all the discrete line intensities.

The beam cannot be truly Gaussian since the Gaussian distribution is continuous from $-\infty$ to $+\infty$. Amplifiers may reasonably reproduce the Gaussian beam profile within the $-3\sigma_{SPOT}$ to $+3\sigma_{SPOT}$ limits or a smaller portion. No simple relationship exists between electronic bandwidth and spot size. That is, monitor performance cannot be determined simply by the electronic bandwidth. The bandwidth number quoted depends on the manufacturer's philosophy on rise time issues. Monitors with the same electronic bandwidth may have different resolutions.

7.4. PIXELS

A pixel is the smallest imaginary area that can exist on a display. The pixel size indicates how many individual pieces of information the system can carry. It does not indicate the extent to which the detail will be resolved. Adjoining pixels can be displayed with different intensities. Here, pixel size is related to resolution. However, displayed pixels can be overlapping and the resolution may be much poorer than that suggested by the pixel size. This is further discussed in Section 7.6., *Addressability*.

The vertical pixel size is typically assumed to equal the line width. It is the display height divided by the number of active scan lines. The number of raster lines limits vertical resolution. This is dictated by the video standard.

182 CCD ARRAYS, CAMERAS, and DISPLAYS

In the horizontal direction, the maximum number of pixels is limited by the spot size. With an optimum system, the electron beam diameter and video bandwidth are matched. For television receivers, the number of monitor pixels is matched to the video standard. These values were given in Table 5-2, page 129.

The definition of a pixel for color systems is more complicated. The representation shown in Figure 7-3 (page 177) implies that the electron beam passes through one hole only. Due to alignment difficulties between the mask and phosphor dots and because the phosphor may have imperfections, CRTs have electron beams that encompass several shadow mask holes. The beam width at the 5% intensity level typically illuminates about 2.5 holes. Encompassing a non-integral number of holes minimizes aliasing and color Moiré patterns[10]. The shadow mask samples the continuous electron beam to create a sampled output (Figure 7-8). The phasing effects created by the shadow mask are obvious.

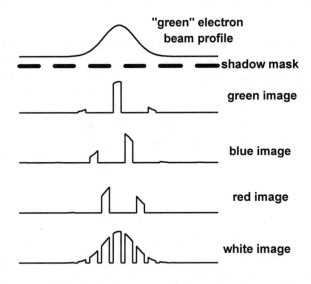

Figure 7-8. Representative image of a single spot on a color CRT. The "green" electron beam is shown, The red and blue beams (not shown) arrive at slightly different angles and provide different intensity profiles. The "sum" appears white when the dot size is less than the eye's resolution limit.

Figure 7-9 illustrates the relationship between the shadow mask pitch and the beam profile. Assume the shadow mask pitch is 0.28 mm. If the 5% diameter covers 2.5 holes, then, using Equation 7-7, $\sigma_{SPOT} = (0.28)(2.5)/4.9 = 0.143$ mm. The spot size (Equation 7-4) is $S = (2.35)(0.14) = 0.336$ mm. In Figure 7-9, the spots are contiguous and each is considered a pixel.

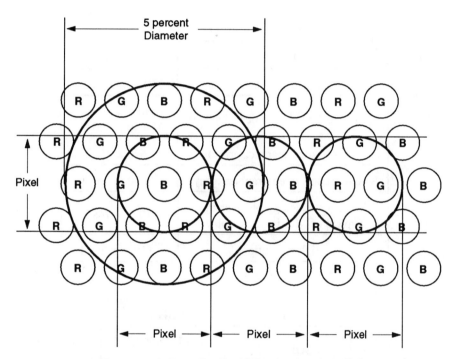

Figure 7-9. Five percent intensity level illuminating 2.5 shadow mask holes. The spot size (FWHM) is equal to the pixel size. If the shadow mask pitch is X mm, then the beam 5% diameter is 2.5X mm, σ_{SPOT} = 0.51X mm, and the pixel is 1.2X mm.

The pixel format is the arrangement of pixels into horizontal and vertical rows. Table 7-4 lists the common formats found in computer color monitors. The format represents the CRT controller's ability to place the electron beam in a particular location. The CRT is an analog device even though the format is listed as discrete values. These monitors are said to be digitally addressed. The pixel format is (erroneously) called the resolution. The relationship between resolution and format is discussed in Section 7.6., *Addressability*.

184 CCD ARRAYS, CAMERAS, and DISPLAYS

Pixel density is usually specified in units of dots per inch. Table 7-5 provides the approximate relationship between "resolution" and pixel density. Confusion exists because dots also refer to the individual phosphor sites. In this context, a "dot" is a pixel. The horizontal pixel density is the number of horizontal pixels (Table 7-4) divided by the horizontal display size.

Table 7-4
FORMAT OF COLOR DISPLAYS

TYPE	FORMAT
CGA (color graphics card adapter)	340 x 200 pixels
EGA (extended graphics adapter)	640 x 350 and 640 x 400
VGA (video graphics array)	640 x 480
SVGA (super video graphics array)	800 x 600
XGA (extended graphics array) Typical of Multi-Sync displays	1024 x 768
High resolution	>1024 x 1024

Table 7-5
TYPICAL PIXEL DENSITY

"RESOLUTION"	PIXEL DENSITY (dpi)
Ultra-high	> 120
High	71 - 120
Medium	50 - 70
Low	< 50

CRT-BASED DISPLAYS 185

Figure 7-10 illustrates the "resolution" of three displays. For a fixed format, as the display size increases, the pixel size also increases to yield a decreasing pixel density. Plotted also is the human eye's resolution capability of 1 arc minute per line expressed at a viewing distance three times the display height. The 1280 x 960 display provides about the highest perceivable resolution. A higher pixel density will be beyond the eye's resolution limit and therefore ineffective. While consumer receivers are viewed at 4.2 times the display height, computer monitors are viewed at a closer distance. Figure 7-10 illustrates that the observer is an integral component of perceived resolution. When monitors are designed, a nominal viewing distance is assumed.

The user is typically interested in performance and not in specific design specifications. For example, the dot pitch is typically superfluous information to the user. Pixel size and the display modulation transfer function (discussed in Section 10.6., *Display*) are usually more important. The *system* pixel is defined as the larger of: (a) the camera pixel, (b) the size limited by the video bandwidth, or (c) the area limited by the CRT spot diameter. The actual pixel size is the convolution of all three.

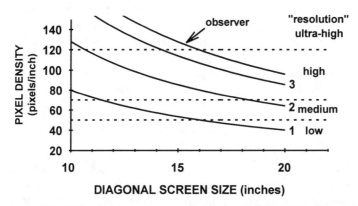

Figure 7-10. "Resolution" of three displays: (1) 640 x 480, (2) 1024 x 768, and (3) 1280 x 960. The observer's resolution is plotted for a viewing ratio of three. The 1280 x 960 display is near the observer's ability to see the pixels.

186 CCD ARRAYS, CAMERAS, and DISPLAYS

7.5. RESOLUTION

Resolution metrics are intertwined with television standards such that they are often presented together. Nearly all televisions are built to the same basic design so that by specifying EIA 170 or NTSC, the television receiver bandwidth and spot size is implied. In this sense, a video standard has a "resolution."

Resolution is independent of display size. Displays can be built to any size and the viewing distance affects the *perceived* resolution. For example, if a screen is designed for viewing at 45 inches, then it may have a 10-inch diagonal (see Figure 7-1, page 173). If viewing will be at 90 inches, the entire display can simply be made larger (20-inch diagonal). The larger display has larger raster lines. But the video electronic bandwidth is the same and the resolution remains constant. When viewed at 45 inches, the larger display will appear to have poorer image quality.

7.5.1. VERTICAL RESOLUTION

The vertical resolution is limited by the number of raster lines. For example, NTSC compatible monitors display 485 lines. However, to see precisely this number, the test pattern must be perfectly aligned with the raster pattern. Since the test pattern may be randomly placed, the pattern at Nyquist frequency can have a zero output if 180° out-of-phase (discussed in Section 8.3.3., *Image Distortion*). To account for random phases, the Kell factor[11] is applied to the vertical resolution to provide an average value. A value of 0.7 is widely used:

$$R_{VERTICAL} = (active\,scan\,lines)(Kell\,factor) \qquad (7\text{-}8)$$

Timing considerations force the vertical resolution to be proportional to the number of vertical lines in the video signal. Therefore, NTSC displays have an average vertical resolution of (0.7)(485) = 340 lines. PAL and SECAM displays offer an average of (0.7)(575) = 402 lines of resolution.

7.5.2. THEORETICAL HORIZONTAL RESOLUTION

While the number of horizontal pixels was given in Table 5-2, page 129, the theoretical horizontal resolution is usually normalized to the picture height. If the video electronics has a band width of f_{VIDEO}, then theoretical number is

$$R_{TVL} = \frac{(2f_{VIDEO})(active\ line\ time)}{aspect\ ratio} \quad \frac{TVL}{PH} \quad (7\text{-}9)$$

The units are television lines/picture height TVL/PH. Unfortunately, careless usage results in using only "lines." Note that there are two TV lines per cycle. While the video band width can be any value, for consumer applications, it is matched to the video standard. For NTSC television receivers, the nominal horizontal resolution is (2)(4.2 MHz)(52.45 µs)/(4/3) = 330 TVL/PH. PAL and SECAM values are approximately 426 and 465 TVL/PH respectively. These values change slightly depending on the active line time and bandwidth selected.

Displays designed for scientific applications usually require the standard video format timing signal but have a much wider video bandwidth. These monitors can display more horizontal TV lines per picture height (TVL/PH) than that suggested by, say, the NTSC video transmission bandwidth. The horizontal bandwidth can be any value and therefore the theoretical horizontal resolution can be any value.

7.5.3. TV LIMITING RESOLUTION

Whereas the flat field condition exists when two adjacent lines cannot be resolved, the TV limiting resolution is a measure of when alternate vertical bars are just visible (on-off-on-off). The standard resolution test target is a wedge pattern with spatial frequency increasing toward the apex of the wedge. It is equivalent to a variable square wave pattern.

The measurement is a perceptual one and the results vary across the observer population. The industry selected[8] the limiting resolution as a bar spacing of $1.18\sigma_{SPOT}$. This result is consistent with the flat field condition. The flat field condition was determined for two Gaussian beams whereas square waves are used for the TV limiting resolution test. Suppose the Gaussian beams in Figure 7-7 (page 181) were separated by an imaginary "black" beam. Then the beams would be separated by just σ_{SPOT} but the intensity distribution would be identical to Figure 7-7. The black beam does not contribute to the visible image.

188 CCD ARRAYS, CAMERAS, and DISPLAYS

For TV limiting resolution, the "on" lines are separated by $(2)(1.18\,\sigma_{SPOT})$. Because bar targets are used, the resultant image is not a precise Gaussian beam and a larger line separation is required ($1.18\sigma_{SPOT}$ versus σ_{SPOT}) for the TV limiting resolution.

The flat field condition and high TV limiting resolution are conflicting requirements. For high TV limiting (horizontal) resolution, σ_{SPOT} must be small. On the other hand, raster pattern visibility (more precisely, invisibility) suggests that σ_{SPOT} should be large. This is further discussed in Section 7.6., *Addressability*.

Example 7-1
SPOT SIZE versus TV LIMITING RESOLUTION

Consider a 14-inch diagonal monitor used to display an image encoded in the EIA 170 format. Assume the raster line spacing is $2\sigma_{SPOT}$. What is the horizontal resolution?

The vertical extent of the 4:3 aspect ratio monitor is 8.4". With 485 lines, each line is separated by 0.0173". σ_{SPOT} is $0.0173/2 = 0.00865$" and the spot size (FWHM) is 0.0204".

EIA 170 contains approximately 440 elements in the horizontal direction. The display width is 11.2 inches and each element width is $11.2/440 = 0.0254$". On the other hand, TV limiting resolution suggests that a bar width of $(1.18)(0.0865) = 0.102$" can be perceived. Since the element width due to the video bandwidth is narrower than the TV limiting resolution, resolution is limited by the spot size. That is, resolution is not limited by the EIA 170 bandwidth.

7.5.4. MTF

TV limiting resolution occurs when the bar spacing is $1.18\sigma_{SPOT}$. Since a cycle consists of two TV lines, the square wave fundamental frequency is $1/(2.35\sigma_{SPOT})$ and the MTF is 0.028. Conversely, the display resolution is selected as that spatial frequency that provides an MTF of 0.028. The MTF equation is found in Section 10.6., *Displays*. Any spatial frequency could be used to specify resolution and the slightly different values are cited in the literature. For commercial applications, the resolution is often specified by the spatial frequency where the MTF is 10%.

On the other hand, square waves are the most popular test patterns. A specification could be: "The CTF must be greater than 10% when viewing 320 line pairs (640 pixels in an on-off-on-off configuration)." Note that by using square wave targets, the system response is the contrast transfer function and not the MTF (discussed in Section 9.4., *Contrast Transfer Function*). Careless usage results in calling the system response to square waves, the MTF. This makes it extremely difficult to compare specifications or calculate performance from published specifications.

When designing a simulation system, the distance to the display and display size are selected first. These are based on physical constraints of the overall system design. The simulation industry generally accepts resolution as that spatial frequency (line pair) which provides a CTF of 10% at this viewing ratio. Since the number of lines depends on what is seen visually, the simulation industry uses units of optical line pairs (OLP). This translates into arc min per line pair.

7.6. ADDRESSABILITY

Addressability is a characteristic of the display controller and represents the ability to select and activate a unique area on the screen. It defines how precisely an electron beam can be positioned on the CRT screen. Addressability is given as the number of discrete lines per picture height or pixels per unit length. With this definition, the inverse of addressability is the center-to-center spacing between adjacent pixels. Addressability is the image format discussed in the Section 7.4., *Pixels* (page 181) and provided in Table 7-4 (page 184).

Addressability and resolution are independent of each other. If the resolution is low, successive lines will over write proceeding lines. If addressability is low, adjacent raster lines will not merge and they will appear as stripes. This was illustrated in Figure 7-7 (page 181) where the line spacing is inverse of the addressability and resolution is the inverse of σ_{SPOT}. Figure 7-7 illustrates good addressability (lines can be placed close together) but poor resolution (cannot detect lines because the modulation is low).

There are two opposing design requirements. The first design requirement is that the raster pattern be imperceptible to the observer. This is the called adjacent pixel requirement. If the display meets this requirement, the picture will appear uniform and solid. It provides the flat field condition. Alphanumeric characters will appear well constructed and highly legible. The second design requirement is the alternating pixel requirement. Here, individual lines, one pixel on and one pixel off should be highly visible.

Increased addressability favors the adjacent pixel requirement. Placing large spots close together will eliminate the visibility of the raster. But this also reduces the visibility of alternating pixels. Similarly an increase in resolution favors the alternating requirement but may make the raster pattern visible.

The resolution/addressability ratio[12] (RAR) is the ratio of FWHM spot size to pixel spacing, P:

$$RAR = \frac{S}{P} = \frac{2.36 \, \sigma_{SPOT}}{P} \qquad (7\text{-}10)$$

P is the (monitor) pixel pitch. For example, if a display is 27.5 cm high and it displays (addresses) 1024 lines, then P is 27.5 cm/1024 lines = 0.27 mm/line. Assuming a 0.28 mm-wide spot then RAR is 0.28/0.27 = 1.04.

As shown in Figure 7-7 (page 181), the summation of periodically spaced spots creates a modulated output. Modulation is defined as $(V_{MAX} - V_{MIN})/(V_{MAX} + V_{MIN})$ where V_{MAX} and V_{MIN} are the maximum and minimum values in the output. Figure 7-11 illustrates the adjacent pixel (on-on-on-on) and alternating pixel (on-off-on-off) modulation as a function of the RAR. An RAR of 1.0 is considered desirable. In Figure 7-7, the RAR is 1.18.

Before the RAR was defined, display design was based on the experimentally determined flat field condition. The flat condition occurs when $P = 2\sigma_{SPOT}$ or RAR = 1.18. Here, the adjacent pixel modulation is near zero - the lines are not visible. The TV limiting resolution describes the alternating pixel situation. Now, $P = 1.18\sigma_{SPOT}$, RAR = 2 and the modulation is near zero (e.g., the lines are not discernible).

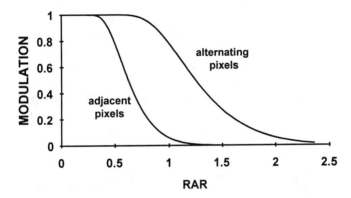

Figure 7-11. Modulation based on Gaussian beam profiles for the adjacent pixel and alternating pixels. An RAR of one is considered desirable. The modulation of the alternating pixel is associated with the display's Nyquist frequency. The modulation of the adjacent pixel is associated with the display's sampling frequency. These are imaginary frequencies since the CRT is an analog device. They represent the ability to place a spot at the desired location on the CRT face plate.

Note that the modulation transfer function is traditionally defined for sinusoidal inputs. The residual ripple is not truly sinusoidal and therefore the modulation is not exactly what would be expected if the input was sinusoidal. That is, the modulation shown in Figure 7-11 is not a true MTF. The measured display response may provide different values than those shown in Figure 7-11.

The alternating pixel modulation represents one Gaussian beam on and one off. This is different from one bar on and one bar off. Resolution tests are performed with square waves. For an analog system, the contrast transfer function is $4/\pi$ times greater than the MTF (discussed in Section 9.4., *Contrast Transfer Function*).

If the horizontal format supports 640 pixels (see Table 7-4, page 184), then 320 cycles should be perceived. But the modulation is not specified at this resolution. For high-quality monitors, the 320 cycles are easily visible (RAR < 1). With average monitors, the 320 cycles are visible but with reduced modulation (RAR ≈ 1). On a poorly designed monitor, the cycles will be barely perceived (RAR > 1).

192 CCD ARRAYS, CAMERAS, and DISPLAYS

While any pixel format can be specified, the RAR determines if the pixels are visible. For the configuration shown in Figure 7-9, (page 183), if the shadow mask pitch is 0.28 mm, then the pixels will be separated by 0.336 mm when the RAR is one. If using a 13-inch diagonal display, the width is 264 mm and 786 horizontal pixels are viewable when the RAR = 1. More pixels can be displayed, but with reduced modulation (Table 7-6 and Figure 7-12). Thus the pixel format, by itself, does not fully characterize the display capability. For optimized displays, it is reasonable to assume that the RAR is about one.

Table 7-6
ALTERNATING PIXEL MODULATION
13-INCH DIAGONAL DISPLAY and 0.28 mm shadow mask pitch
5% intensity level illuminating 2.5 shadow mask holes

PIXELS	RAR	MODULATION
800	1.02	0.75
900	1.15	0.61
1000	1.27	0.47
1100	1.40	0.35
1200	1.53	0.25

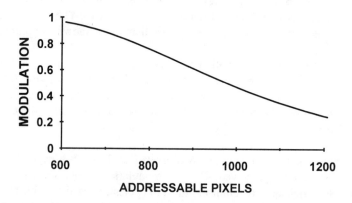

Figure 7-12. 13-inch diagonal display alternating pixel (Gaussian beam) modulation. The shadow mask pitch is 0.28 mm. The 5 percent intensity level illuminates 2.5 shadow mask holes.

ANSI/HFS 100-1988 requires that the adjacent pixel modulation be less than 20%. This deviates significantly from the flat field condition (adjacent pixel modulation is 0.014). ISO 9241 recommends that the alternating pixel modulation be greater than 0.4 for monochrome and greater than 0.7 for color systems. For monochrome systems, the RAR will vary from 0.8 to 1.31. For color systems the RAR varies from 0.8 to 1.04. These standards were developed to assure legibility of alphanumeric characters.

If the RAR is too high, then the output of two adjacent "on" lines is much greater then the output of two separated lines due to summing effect (Figure 7-13). For the flat field condition (Figure 7-7, page 181), the RAR is 1.18 and the summed lines are 27% brighter than an individual line. If the RAR is very large, the intensity of several adjacent lines on can be significantly brighter than a single line on. Thus, there is also a tradeoff in intensity with RAR.

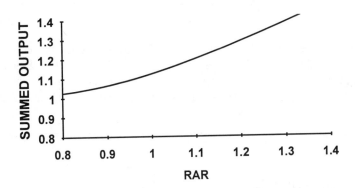

Figure 7-13. Summed output of contiguous lines on. Ideally, the intensity should remain constant whether one, two, or multiple lines are on.

Multi-sync monitors automatically adjust the internal line rate to the video line rate. These monitors can display any format from 525 lines up to 1225 lines. Since the displayed image size remains constant, the vertical line spacing changes. At the 1225 line format, the raster lines are close together and form a uniform (flat) field. At the lower line rates, the individual raster lines are separated and may be perceived. Since the same video amplifiers are used for all line rates, the horizontal resolution is independent of the line rate. That is, the RAR in the horizontal direction stays constant whereas the RAR in the vertical direction changes with the number of lines. If the vertical RAR is one at 1225 lines, then it must be $525/125 = 0.43$ at 525 lines.

Example 7-2
DISPLAY RESOLUTION

A multi-sync monitor has a horizontal resolution of 800 TV lines per picture height. If the display diagonal is 14", what is the spot size? Are the raster lines visible?

For a 4:3 aspect ratio display, the CRT vertical extent is 284 mm. The estimated spot size is 284/800 = 0.355 mm and σ_{SPOT} is S/2.35 = 0.335/2.35 = 0.15 mm. If the number of displayed lines is N_{LINE}, the RAR is

$$RAR = \frac{S N_{LINE}}{H_{MONITOR}} = \frac{0.355}{284} N_{LINE} = \frac{N_{LINE}}{800} \quad (7\text{-}11)$$

With 485 displayed lines (EIA 170), the RAR is 0.61. From Figure 7-11 the raster pattern (adjacent pixels) is quite visible. For EIA 343A with 809 active lines, the RAR is at the desired value of one. At 946 active lines, the RAR increases to 1.18. Here the alternating pixel modulation starts to drop and the vertical resolution starts to decrease.

7.7. SHADES OF GRAY

An often used, but widely misunderstood, concept is "shades of gray." While displays may be specified in shades of gray where each step is square root of 2, human vision cannot. The misunderstanding originated with the thought that the eye requires a $\sqrt{2}$ increment to just detect an increment in luminance. Studies of visual perception clearly indicate that just noticeable increments in luminance vary with background luminance level, size and shape of the target, the ambient luminance, and a variety of other variables. To use a single number to characterize the human eye is too simplistic. The $\sqrt{2}$ approach is merely another way to indicate display dynamic range. No other inference should be made about the definition.

CRT-BASED DISPLAYS   195

7.8. CHARACTER RECOGNITION

Computer monitors are designed to for maximum legibility of alphanumeric characters and enhanced graphical capability. Readable characters can be formed in any block of pixels from 5 x 7 to 9 x 16 and still be aesthetically pleasing.

The ability to see detail is related to the RAR. Figure 7-14 illustrates three letters and the resultant intensity traces. The RAR must be near one so that a reasonable contrast ratio exits between on and off pixels. With reasonable contrast, the inner detail of the character is seen and the character is legible. Similarly, with a reasonable contrast ratio, adjacent letters will appear as separate letters. With characters, the alternating pixel pattern is called one stroke separated by a space.

Figure 7-15 illustrates modulation affects character legibility. For easy interpretation, the system contains a shadow mask but only white and black can be seen. The image is magnified so that the individual phosphor dots can be seen. The sampled nature of the Gaussian electron beam is evident. As the beam width increases, the modulation decreases and the RAR increases. The characters become blurry as the RAR increases.

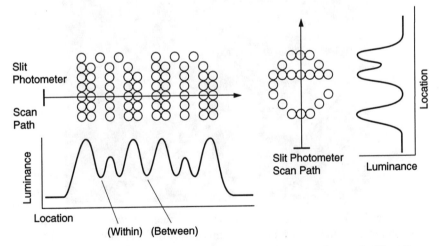

Figure 7-14. The RAR must be near one so that the inner details of characters are visible and the double width appears solid. Similarly with RAR near one, two adjacent characters appear separate. Each dot represents a pixel. Lower RARs will make the characters appear very sharp. However, the raster pattern will become visible and the characters will not appear solid.

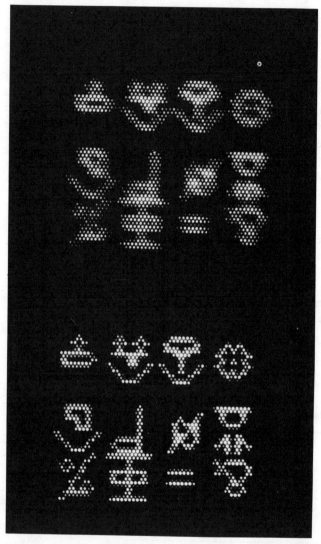

Figure 7-15. Complex symbols are more legible when the display modulation is high. Each dot represents a hole in the shadow mask. Top: modulation = 0.5 at the Nyquist frequency. Bottom: modulation = 0.1 at the Nyquist frequency. When viewed at 10 feet, the individual dots are no longer perceptible. At this distance, the top characters appear sharp (RAR ≈ 1.25) and the bottom ones appear fuzzy (RAR ≈ 1.8). By courtesy of the Mitre Corporation.

7.9. CONTRAST

Selection of one minute of arc as a measure of human visual acuity provided back-of-the-envelope calculations for selecting pixel size, display size, and viewing distance. However, the eye's response varies according to the displayed contrast.

The air-glass interface and the phosphor reflect ambient light. This reduces contrast between pixels and reduces the display's color gamut. As a result, most displays are measured in a darkened room where the reflected light is less than 1 percent of the projected light. Typically the contrast is measured with a checkerboard pattern consisting of sixteen alternating rectangles. The average value of the eight white rectangles and the average value of the eight black rectangles are used to calculate the contrast ratio.

The perceived contrast ratio is

$$Contrast\ ratio = \frac{L_{WHITE} + L_{AMBIENT} \cdot reflectivity}{L_{BLACK} + L_{AMBIENT} \cdot reflectivity} \quad (7\text{-}12)$$

L_{WHITE} and L_{BLACK} represent the luminances from the white and black rectangles respectively. The reflections are the same regardless if a pixel is on or off. About 4 percent of the ambient light, $L_{AMBIENT}$, is reflected at the air-glass interface. The CRT phosphor is an excellent Lambertian reflector with a typical reflectivity of 70%. "Black" displays minimize these reflections and thereby increase the perceived contrast.

Unfortunately, specifications that are based on no ambient lighting can only be achieved in a totally darkened room. Safety considerations require some lighting in all rooms so that the specification can never be achieved in practice.

The contrast ratio for typical displays is about 20:1 under normal ambient lighting conditions. This decreases as the ambient light increases. The eye operates in the reverse way. As the ambient lighting increases, more contrast is necessary to see target-background differences. Thus as the displayed contrast decreases in bright light, the eye requires more contrast. This explains why it is so difficult to see television imagery in sunlight and why it appears so bright in a darkened room.

7.10. REFERENCES

1. L. E. Tannas Jr., ed., *Flat panel Displays and CRTs*, Van Nostrand Reinhold Co. (1985).
2. J. A. Castellano, *Handbook of Display Technology*, Academic Press (1992).
3. S. Sherr, *Electronic Displays*, 2nd edition, Wiley and Sons (1993).
4. J. C. Whitaker, *Electronic Displays*, McGraw-Hill (1994).
5. J. Peddie, *High-Resolution Graphics Display Systems*, Windcrest/McGraw-Hill, NY (1994).
6. T. Olson, "Behind Gamma's Disguise," *SMPTE Journal*, Vol. 104(7), pp. 452-458 (1995).
7. W. Cowan, "Displays for Vision Research," in *Handbook of Optics*, M. Bass, ed., 2nd edition, Vol. 1, Chapter 27, McGraw-Hill, New York, NY (1995).
8. L. M. Biberman, "Image Quality," in *Perception of Displayed Information*, L. M. Biberman, ed., pp. 13-18: Plenum Press, New York (1973).
9. P. G. J. Barten, "Spot Size and Current Density Distributions of CRTs," in Proceedings of the SID, Vol. 25(3), pp. 155-159 (1984).
10. A. Miller, "Suppression of Aliasing Artifacts in Digitally Addressed Shadow-mask CRTs," *Journal of the Society of Information Display*, Vol. 3(3), pp 105-108 (1995).
11. S. C. Hsu, "The Kell Factor: Past and Present," *SMPTE Journal*, Vol. 95, pp. 206-214 (1986).
12. G. M. Murch and R. J. Beaton, "Matching Display Resolution and Addressability to Human Visual Capacity," *Displays*, Vol. 9, pp. 23-26 (Jan 1988).

8

SAMPLING THEORY

Sampling (digitization) is an inherent feature of all imaging systems. The scene is spatially sampled in both directions due to the discrete locations of the CCD detector elements. This sampling creates ambiguity in target edges and produces Moiré patterns when viewing periodic targets. Aliasing becomes obvious when (a) the image size approaches the detector size and (b) the detectors are in a periodic lattice. Spatial aliasing is rarely seen in photographs or motion pictures because the pigments are randomly dispersed.

If the detector size and spacings are different in the horizontal and vertical directions, sampling effects will be different in the two directions. This leads to Moiré patterns that are a function of the target orientation with respect to the CCD array axis. While sampling is two-dimensional, for simplicity, sampling effects will be presented in one-dimension.

The highest frequency that can be faithfully reconstructed is one-half the sampling rate. Any input signal above the Nyquist frequency, f_N, (which is defined as one-half the sampling frequency, f_S) will be aliased down to a lower frequency. That is, an undersampled signal will appear as a lower frequency after reconstruction (Figure 8-1).

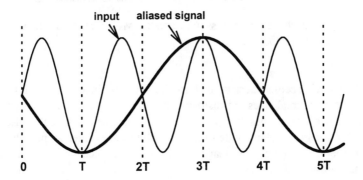

Figure 8-1. An undersampled sinusoid will appear as a lower frequency after reconstruction. The sampling frequency is $f_S = 1/T$. When T is the measured in mrad, mm, or time, the sampling frequency is expressed in cy/mrad, cy/mm, or Hz respectively.

200 CCD ARRAYS, CAMERAS, and DISPLAYS

Signals can be undersampled or oversampled. Undersampling is a term used to denote that the input frequency is greater than the Nyquist frequency. It does not imply that the sampling rate is inadequate for any specific application. Similarly, oversampling does not imply that there is excessive sampling. It simply means that there are more samples available than that required by the Nyquist criterion.

After aliasing, the original signal can never be recovered. Undersampling creates Moiré patterns (Figure 8-2). Diagonal lines appear to have jagged edges or "jaggies." Aliasing is not always obvious when viewing complex scenery and, as such, is rarely reported during actual system usage although it is always present. It may become apparent when viewing periodic targets such as test patterns, picket fences, plowed fields, railroad tracks, and Venetian blinds.

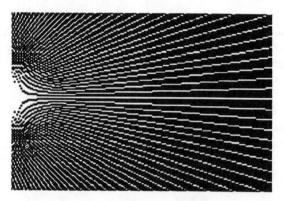

Figure 8-2. A raster scan system creates Moiré patterns when viewing wedges or star bursts.

We have become accustomed to the aliasing in commercial televisions. Periodic horizontal lines are distorted due to the raster. Cloth patterns such as herringbones and stripes also produce Moiré patterns. Cross-color effects occur in color imagery. Many video tape recordings are undersampled to keep the price modest and yet the imagery is considered acceptable when viewed at normal viewing distances.

While aliasing is never considered desirable, it is accepted in scientific and military monochrome applications where the modulation transfer function (MTF) must be maintained. High MTF is related to image sharpness. Aliasing becomes bothersome when the scene geometric properties must be maintained as with mapping. It affects the performance of most image-processing algorithms.

SAMPLING THEORY 201

Aliasing is reduced to an acceptable level in consumer color sensors. For specific applications such as medical imaging, it may be considered unacceptable. In medical imagery, an aliased signal may be misinterpreted as a medical abnormality that requires medication, hospitalization, or surgery.

Sampling theory states that the *frequency* can be unambiguously recovered for all input frequencies below Nyquist frequency. It was developed for band-limited electrical circuits. The extension to imaging systems is straight forward. However, the eye is primarily sensitive to intensity variations. Sampling theory states nothing about the intensity nor the appearance of the signal. Furthermore, sampling theory deals with sinusoids whereas the real world contains all frequencies.

Sampling theory is traditionally presented from a modulation transfer function view point. MTF theory is presented in Chapter 9 and the subsystem MTFs are discussed in Chapter 10. Linear system theory is used for system analysis because of the wealth of mathematical tools available. Approximations are used to account for sampling effects. While these approximations are adequate to describe the MTFs, they do not indicate how imagery may be distorted.

Distortion is most obvious with periodic structures. These generally do not exist in nature and aliasing is rarely reported when viewing natural scenery. Aliasing may be pronounced with manufactured objects (e.g., test targets, picket fences, plowed fields, etc.). Distortion effects can only be analyzed on a case-by-case basis. Since the aliasing occurs at the detector, the signal must be band-limited by the optical system to prevent it.

Sampled data systems are nonlinear and do not have a unique MTF[1-6]. The "MTF" depends on the phase relationships of the scene with the sampling lattice. Superposition does not hold and any MTF derived for the sampler cannot, in principle, be used to predict results for general scenery. To account for the nonlinear sampling process, a sample-scene MTF is added (discussed in Section 10.9., *Sample Effects*). As the sampling rate increases, the MTF becomes better defined. As the sampling rate approaches infinity, the system becomes an analog system and the MTF is well defined. Equivalently, as the detector size and detector pitch decrease, signal fidelity increases.

The symbols used in this book are summarized in the *Symbol List* (page xiv) and it appears after the *Table of Contents*.

8.1. SAMPLING THEOREM

In a sampled-data system, the sampling frequency interacts with the signal to create sum and difference frequencies. Any input frequency, f, will appear as $nf_S \pm f$ after sampling ($n = -\infty$ to $+\infty$). Figure 8-3 illustrates a band-limited system with frequency components replicated by the sampling process. The base band ($-f_H$ to f_H) is replicated at nf_S. To avoid distortion, the lowest possible sampling frequency is that value where the base band adjoins the first side band (Figure 8-3c). This leads to the sampling theorem that a band-limited system must be sampled at twice the highest frequency ($f_S \geq 2f_H$) to avoid distortion in the reconstructed image. The sampling theorem applies to any periodic input frequency. The units can be either spatial frequency or electrical frequency.

After digitization, the data resides in data arrays (e.g., computer memory location) with nonspecific units. The user assigns units to the arrays during image reconstruction. That is, the data is read out of the memory in a manner consistent with display requirements. Since monitors are analog devices, the data is transformed into an analog signal using a sample-and-hold circuit. This creates a "blocky" image and the individual blocks are labeled as pixels on a screen. Moiré patterns only exist after the image is presented to an observer. They do not exist in the sampled data. The "blockiness" is removed by a post-reconstruction filter (discussed in Section 10.5.3., *Post-Reconstruction Filter*). If the original signal was oversampled and if the post-reconstruction filter limits frequencies to f_N, then the reconstructed image can be identical to the original image.

8.2. ALIASING

As the sampling frequency decreases, the first side band starts to overlap the base band and the power spectrums add (Figure 8-4). The overlaid region creates distortion in the reconstructed image. This is aliasing. Once aliasing has occurred, it cannot be removed. Within an overlapping band, there is an ambiguity in frequency. It is impossible to tell whether the reconstructed frequency resulted from an input frequency of f or $nf_S \pm f$.

All frequency components above f_N are folded back into the base band so that the base band contains

$$O_{BASE-BAND} = \sum_{n=-\infty}^{\infty} O(nf_S \pm f) \quad \text{where } nf_S \pm f \leq f_N \qquad (8\text{-}1)$$

SAMPLING THEORY 203

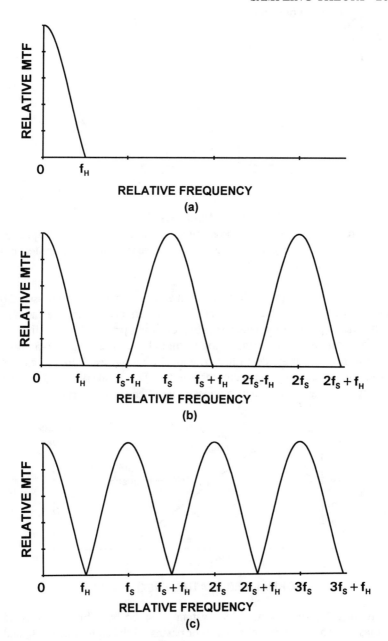

Figure 8-3. Sampling replicated frequencies at $nf_S \pm f$. (a) Original band-limited signal, (b) frequency spectrum after sampling, and (c) when $f_S = 2f_H$, the bands just adjoin.

204 CCD ARRAYS, CAMERAS, and DISPLAYS

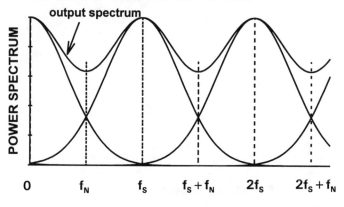

Figure 8-4. Aliasing alters both the signal and noise baseband spectra. In a real system, signals and noise are not necessarily band-limited and some aliasing occurs.

To avoid aliasing in the electronic domain, the signal may be passed though a low-pass filter (anti-aliasing filter). These filters cannot remove the aliasing that has taken place at the detector. They can only prevent further aliasing that might occur in the downstream analog-to-digital converter. Figure 8-5 illustrates the cutoff features of an ideal filter. The ideal filter passes all frequencies below f_N and attenuates all frequencies above f_N. The ideal filter is, of course, unrealizable. It also can produce undesirable effects such as ringing.

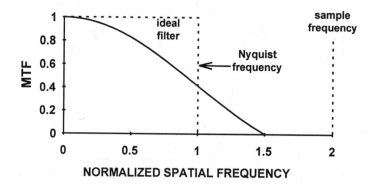

Figure 8-5. An undersampled system must be band-limited to avoid aliasing. The ideal anti-aliasing filter passes all the signals below f_N and no signal above f_N. The ideal filter is unrealizable. A very sharp cutoff filter may introduce ringing in the output. Whether aliasing is considerable undesirable depends on the application.

SAMPLING THEORY 205

Optical band-limiting can be achieved by using small diameter optics, defocusing, blurring the image, or by inserting a birefringent crystal between the lens and array. The birefringent crystal changes the effective sampling rate and is found in almost all single chip color cameras (discussed Section 10.3.4., *Optical Anti-Alias Filter*). Unfortunately, these approaches also degrade the MTF (reduce image sharpness) in the base band and typically are considered undesirable for scientific applications. For medical applications, where aliased imagery may be misinterpreted as an abnormality, reduced aliasing usually out weighs any MTF degradation.

According to the sampling theorem, any sampled frequency, f, where $f \leq f_N$, can be uniquely recovered. However, the square wave is the most popular test target and it is characterized by its fundamental frequency, f_o. When expanded into a Fourier series, the square wave consists of an infinite number of frequencies (discussed in Section 9.4., *Contrast Transfer Function*). Although the square wave fundamental may be oversampled, the higher harmonics will not. The appearance of the square wave after reconstruction depends on the relative values of the optical, detector, and electronic MTFs.

Although no universal method exits for quantifying aliasing, it is reasonable to assume that it is proportional to the MTF that exists above f_N. In Figure 8-6 a significant MTF exists above the Nyquist frequency. Whether this aliasing is objectionable, depends on the camera application.

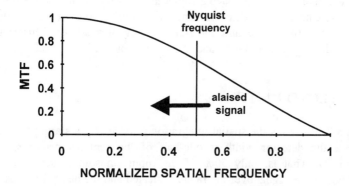

Figure 8-6. Signals above Nyquist frequency are aliased down to the base band. The area bounded by the MTF above f_N may be considered as an "aliasing" metric.

If a system is Nyquist frequency limited, then the Nyquist frequency is used as a measure of resolution. Since no frequency can exist above the Nyquist frequency, many researchers represent the MTF as zero above the Nyquist frequency (Figure 8-7). This representation may be too restrictive for modeling purposes.

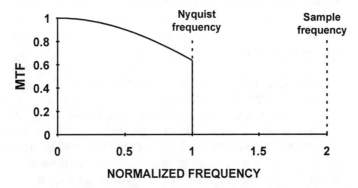

Figure 8-7. MTF representation of an undersampled system as a function of f/f_N. Since no frequency can exist above the Nyquist frequency, some researchers represent the MTF as zero above the Nyquist frequency.

8.3. THE DETECTOR AS A SAMPLER

Figure 8-1 (page 199) illustrated a signal that is sampled at discrete points. It typifies extremely small detectors and flash analog-to-digital converters. Detectors have a finite size and integrate the signal spatially. This spatial integration reduces the modulation and is the detector MTF.

8.3.1. DETECTOR MTF

Figure 8-8 illustrates the spatial integration afforded by a staring array. In Figure 8-8a, the detector width is one-half of the center-to-center spacing (detector pitch). That is, only 50% of the input sinusoid is detected. The detector averages the signal over the range indicated and the heavy lines are the detector outputs. The output MTF is $(V_{MAX} - V_{MIN})/(V_{MAX} + V_{MIN})$ where V_{MAX} and V_{MIN} are the maximum and minimum voltages respectively. As the detector size increases, it integrates over a larger region and the MTF decreases (compare Figure 8-8a with 8-8b). As the phase changes between the detector location and the peak of the sinusoid, the output voltages change. This creates a variable MTF caused by sampling.

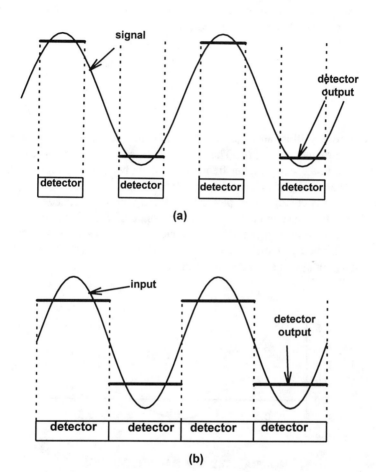

Figure 8-8. A detector spatially integrates the signal. Figure 8-1 (page 199) represents the output of extremely small detectors. (a) Detector width is one-half of the center-to-center spacing and (b) detector width is equal to the center-to-center spacing. (a) is representative of frame transfer arrays and (b) typifies an interline transfer CCD array. The heavy lines are the detector output voltages. As the phase changes, the output also changes.

208 CCD ARRAYS, CAMERAS, and DISPLAYS

The detector MTF cannot exist by itself. Rather, the detector MTF must also have the optical MTF to make a complete imaging system. In one-dimension, the MTF of a single rectangular detector is

$$MTF_{DETECTOR} = \frac{\sin(\pi d f_i)}{\pi d f_i} = sinc(d f_i) \qquad (8\text{-}2)$$

Figure 8-9 illustrates the detector MTF in one dimension. The MTF is equal to zero when $f = k/d$. The first zero ($k=1$) is considered the detector cutoff, f_{DC}, because any higher frequency will not be faithfully reproduced. Any input spatial frequency above detector cutoff will be aliased down to a lower frequency. It is customary to plot the MTF only up to the first zero (Figure 8-10). The optical system cutoff limits the absolute highest spatial frequency that can be faithfully imaged. The detector cutoff may be either higher (optically-limited system) or lower (detector-limited system). Nearly all CCD-based imaging systems are detector-limited.

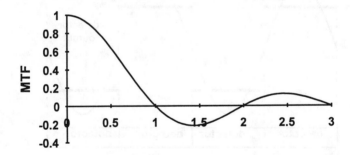

Figure 8-9. Detector MTF as a function of normalized spatial frequency f/f_{DC}. Detector cutoff is $f_{DC} = 1/d$. The negative MTF values represent contrast reversal: periodic dark bars will appear as light bars.

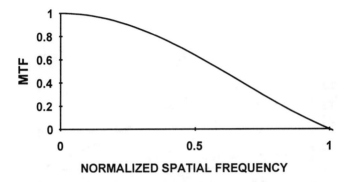

Figure 8-10. Typical detector MTF representation as a function of normalized spatial frequency f/f_{DC}. The MTF is usually plotted only up to the first zero.

8.3.2. DETECTOR ARRAY NYQUIST FREQUENCY

A staring array acts a sampler whose sampling rate is d_{CC} (mm) where d_{CC} is the detector center-to-center spacing (detector pitch). Detector arrays are specified by the fill-factor which is the detector area divided by the cell area. For square detectors with equal spacing in the vertical and horizontal directions, the fill-factor is $(d/d_{CC})^2$. Staring arrays can faithfully reproduce signals up to $f_N = 1/(2d_{CC})$. Although the individual detectors can reproduce higher spatial frequencies, the scene spectrum is sampled at $f_S = 1/d_{CC}$ (Figure 8-11).

Staring arrays are inherently undersampled when compared to the detector spatial frequency cutoff. Any scene spatial frequency above f_N will appear as a lower spatial frequency in the reconstructed image. To avoid aliasing, the scene frequencies must be optically band limited similar to that shown in Figure 8-5 (page 204).

Most single chip color filter arrays have an unequal number of red, green, and blue detectors. A typical array designed for NTSC operation will have 768 detectors in the horizontal direction with 384 detectors sensitive to green, 192 sensitive to red, and 192 sensitive to blue region of the spectrum. Suppose the arrangement is G-B-G-R-G-B-G-R. The "green" detectors will have a center-to-center spacing as shown in Figure 8-11b.

210 CCD ARRAYS, CAMERAS, and DISPLAYS

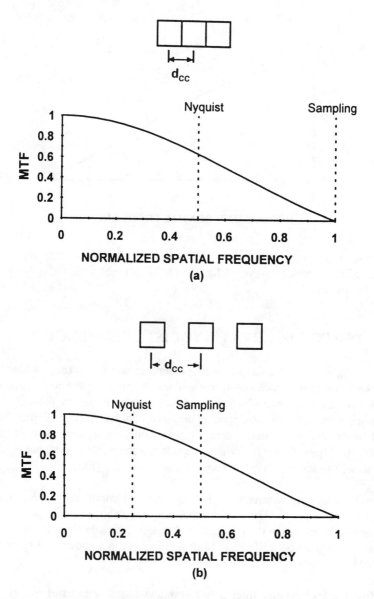

Figure 8-11. Two arrays with different center-to-center spacings. The detector size, d, is the same for both. (a) $d/d_{CC} = 1$. This typifies frame transfer devices. (b) $d/d_{CC} = 0.5$. This is representative of interline transfer CCD arrays. The MTF is plotted as a function of f/f_{DC}. $f_S = 1/d_{CC}$ and $f_N = 1/(2d_{CC})$.

The spacing of the "blue" and "red" detectors is twice the "green" detector spacing. This produces a "blue" and "red" array Nyquist frequency that is one-half of the "green" array Nyquist frequency (Figure 8-12). Other detector layouts will create different array Nyquist frequencies. The "color" Nyquist frequencies can be different in the horizontal and vertical directions. These unequal array Nyquist frequencies create color aliasing in single chip cameras that is wavelength specific. Black-and-white scenes can appear as green, red, or blue imagery[7]. A birefringent crystal, inserted between the lens and the array, changes the effective sampling rate and minimizes color aliasing (discussed Section 10.3.4., *Optical Anti-Alias Filter*).

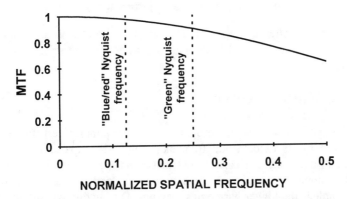

Figure 8-12. Detector MTF normalized to f/f_{DC}. Unequally spaced red, green, and blue sensitive detectors creates different array Nyquist frequencies. This creates different amounts of aliasing and black-and-white scene may break into color. The spacing between the "red" and "blue" detectors is twice the "green" detector spacing.

Aliasing in the vertical direction can be reduced by summing alternate detector outputs (field integration) in a two-field camera system (see Figure 3-18, page 63). Here, the effective detector area is 2d and the cutoff is 1/2d (Figure 8-13). This eliminates aliasing but also reduces the vertical MTF and, therefore, reduces image sharpness in the vertical direction. Horizontal sampling still produces aliasing.

Microlenses increase the effective detector size but do not affect the Nyquist frequency. By increasing the detector area, the detector cutoff decreases. This is equivalent to summing several detectors as illustrated in Figure 8-13. Since each "color" has a different Nyquist frequency, the amount of "summing" may differ by color. Here, each color will have a different MTF and the effect on image quality can only be determined after the camera have been built.

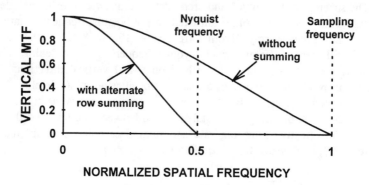

Figure 8-13. Alternate detector summing can reduce aliasing in the vertical direction. Vertical MTF with and without summing normalized to f/f_{DC}.

8.3.3. IMAGE DISTORTION

The square wave (bar pattern) is the most popular test target. It is characterized by its fundamental frequency only. The expansion of a square wave into a Fourier series clearly shows that it consists of an infinite number of sinusoidal frequencies. Although the square wave fundamental frequency may be oversampled, the higher harmonics may not. During digitization, the higher order frequencies will be aliased down to lower frequencies and the square wave will change its appearance. There will be intensity variations from bar-to-bar and the bar width will not remain constant.

The MTF approach only shows how the frequency spectrum is modified by the sampling process. It does not indicate target appearance. Digitization aliases higher order square wave frequencies into the baseband. When two frequencies are close together, they create a beat frequency that is equal to the difference. The fundamental and its replication create a beat frequency of $f_{BEAT} = (f_S - f_o) - f_o$. The beat frequency period lasts for N input frequency cycles:

$$N = \frac{f_o}{2(f_N - f_o)} \qquad (8\text{-}3)$$

Figure 8-14 illustrates N as a function of f_o/f_N. As f_o/f_N approaches one, the beat frequency becomes obvious. Lomheim et. al.[8] created these beat frequencies when viewing multiple bar patterns with a CCD camera.

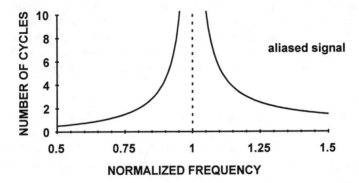

Figure 8-14. Number of input frequency cycles required to see one complete beat frequency cycle as a function of f_o/f_N.

Figures 8-15 through 8-18 illustrate an ideal staring array output when the optical MTF is unity for all spatial frequencies and therefore represents a worst case scenario. That is, all higher harmonics that would have been attenuated by the optical system are available for aliasing by the detector. For these figures, d/d_{CC} is one. The individual detectors sample the square wave in the same manner as illustrated in Figure 8-8b (page 207).

If $f_o/f_N = 0.952$, the beat frequency is equal to 9.9 cycles of the input frequency (Figure 8-15). Here, the target must contain at least 10 cycles to see the entire beat pattern. A 4-bar pattern may either be replicated or nearly disappear depending on the phase.

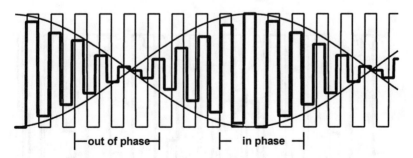

Figure 8-15. Beat frequency produced by an ideal staring system when $f_o/f_N = 0.952$. N = 9.9 and $d/d_{CC} = 1$. The light line is the input and the heavy line is the detector output. The beat frequency envelope is shown.

Standard characterization targets, however, consist of several bars. Therefore the beat pattern may not be seen when viewing a 3-bar or 4-bar target. The image of the bar pattern would have to be moved $\pm\tfrac{1}{2}$ d (detector extent) to change the output from a maximum value (in-phase) to a minimum value (out-of-phase). This can be proven by selecting just four bars in Figure 8-15.

When f_o/f_N is less than about 0.6 (Figure 8-16), the beat frequency is not obvious. Now the output nearly replicates the input but there is some slight variation in pulse width and amplitude. In the region where f_o/f_N is approximately between 0.6 and 0.9, adjacent bar amplitudes are always less than the input amplitude (Figure 8-17).

When f_o/f_N is less than about 0.6, a 4-bar pattern will always be seen (select any four adjoining bars in Figure 8-16). When $f_o/f_N < 0.6$, phasing effects are minimal and a phase adjustment of $\pm\tfrac{1}{2}$d in image space will not affect an observer's ability to resolve a four-bar target. In the region where f_o/f_N is approximately between 0.6 and 0.9, 4-bar targets will never *look* correct (Figure 8-17). One or two bars may be either much wider than the others or one or two bars may be of lower intensity than the others. This phasing effect is a measurement concern and is not normally included in modeling.

Input frequencies of $f_o = f_N/k$, where k is an integer, are faithfully reproduced (i.e., no beat frequencies). When k = 1, as the target moves from in-phase to out-of-phase, the output will vary from a maximum to zero. Selection of f_N/k targets avoids the beat frequency problem but significantly limits the number of spatial frequencies selected (discussed in Section 9.4., *Contrast Transfer Function*).

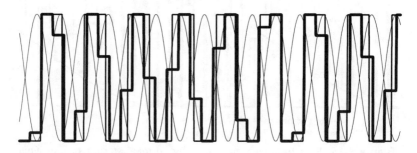

Figure 8-16. Ideal staring system output when $f_o/f_N = 0.522$. N = 0.54 and $d/d_{CC} = 1$. The light line is the input and the heavy line is the detector output. The output nearly replicates the input when $f_o/f_N < 0.6$.

SAMPLING THEORY 215

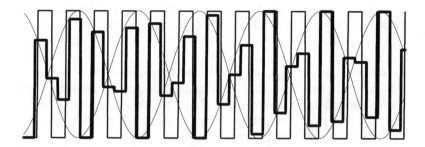

Figure 8-17. Ideal staring system output when $f_o/f_N = 0.811$. N = 2.14. The light line is the input and the heavy line is the detector output. The output never *looks* quite right when f_o/f_N is between 0.6 and 0.9.

Signals whose frequencies are above Nyquist frequency will be aliased down to lower frequencies. This would be evident if an infinitely long periodic target was viewed. However, when f_o is less than about 1.15 f_N, it is possible to select a phase such that four adjoining bars *appear* to be faithfully reproduced (Figure 8-18). These targets can be *resolved* although the underlying fundamental frequency has been aliased to a lower frequency.

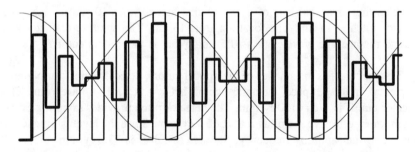

Figure 8-18. Ideal staring system output when $f_o/f_N = 1.094$. N = 5.8. The light line is the input and the heavy line is the detector output. By selecting the appropriate phase, the output *appears* to replicate an input 4-bar pattern.

Figures 8-15 through 8-18 represent the detector output when d/d_{CC} is one. As the d/d_{CC} decreases, the beat frequency becomes less noticeable. It is the finite width detector that emphasizes the beat frequencies. In the limit that the detector is very small, the beat frequency disappears. This was illustrated in Figure 8-1 (page 199).

Systems can *detect* signals whose spatial frequencies are above cutoff but cannot faithfully reproduce them. For example, patterns above the Nyquist frequency are aliased to a frequency below Nyquist and a 4-bar pattern may appear as a distorted 3-bar pattern. If the fundamental frequency is less than about $1.15 f_N$, the pattern may appear as a 4-bar target.

The phasing effects shown in Figures 8-15 through 8-18 "disappear" when the target is moving. Here, each frame provides a different image. For each target whose spatial frequency is above Nyquist frequency, one, two, or three bars are visible depending on the relative phase. As the pattern moves, a different set of bars is visible. The eye blends these individual patterns so that you perceive four bars. This prompted some researchers[9] to develop a dynamic test in which the 4-bar target moves.

Example 8-1
SYSTEM CUTOFF

A staring array consists of detectors that are 10 µm x 10 µm in size. The detector pitch is 15 µm. The focal length is 15 cm. The aperture diameter is 3 cm and the average wavelength is 0.5 µm. What is the system cutoff?

System cutoff is the smaller of the optical cutoff, detector cutoff, or Nyquist frequency. In image space, the optical cutoff is $f_{oc} = D_o/(\lambda fl) = 400$ cy/mm. The detector cutoff is $1/d = 100$ cy/mm. The detector pitch provides sampling every 15 µm for an effective sampling rate at 66.7 cy/mm. Since the Nyquist frequency is one-half the sampling frequency, the system cutoff is 33.3 cy/mm.

Since object space is related to image space by the lens focal length, the optical cutoff is 30 cy/mrad, detector cutoff is 15 cy/mrad, sampling is 10 cy/mrad, and the Nyquist frequency is 5 cy/mrad.

Since the MTF remains zero for all frequencies higher than the optical cutoff, any scene frequency above f_{oc} will appear as a uniform blob with no modulation. Spatial frequencies higher than the detector cutoff or the Nyquist frequency (but less than f_{oc}) will be aliased down to lower frequencies.

8.4. ALIASING in FRAME GRABBERS and DISPLAYS

For many applications, the camera output is an analog signal that is then re-sampled by a frame grabber (frame capture device). If the frame grabber's sampling rate is sufficiently high and the analog signal is band-limited, the reconstructed signal will replicate the analog signal in frequency, amplitude and pulse width.

In principle, the clock rate of an analog-to-digital converter can be set at any rate. However, to conserve on memory requirements and minimize clock rates, some frame grabbers tend to just satisfy the Nyquist frequency of a standard video signal. That is, if the video bandwidth is f_{BW}, then the frame grabber sampling clock operates at $2f_{BW}$ (see Table 5-2, page 129, for f_{BW} values). A frame grabber, with its flash converter, can distort the signal as shown in Figure 8-19. The variation in edge locations affects all image processing algorithms. The variations in pulse width can only be seen on a monitor since the frame grabber data resides in a computer memory. Note that output pulse widths are equally spaced. They just do not match up with the input signal.

Figure 8-19. Frame grabber output after sample and hold circuitry when $f_o/f_N = 0.952$. The light line is the input and the heavy line is the output. Here, f_N is the Nyquist frequency of created by the frame grabber clock rate. The output is representative of what is seen on a monitor.

Most frame grabbers have an internal anti-alias filter. This filter insures that the frame grabber does not produce any additional aliasing. The filter cutoff is linked to the frame grabber clock and is not related to the camera output. Once aliasing has occurred in the camera, the frame grabber anti-alias filter cannot remove it.

Aliasing is rarely discussed with displays, but, under certain conditions can occur. Many displays employ digital techniques to match the electronic signal to the CRT characteristics. Ideally, the internal analog-to-digital converter should have a low-pass filter to eliminate possible aliasing. Although good engineering design recommends that the low-pass filter be included, cost considerations may eliminate it. With standard video formats, the bandwidth is limited by the video format and there is no need for the low-pass filter. If a nonstandard bandwidth is used such as that required by a high resolution CCD camera, these monitors may exhibit some aliasing. This becomes noticeable as the array size increases. For example, a high quality monitor can display 1000 x 1000 pixels. If the array consists of 4000 x 4000 elements, the display may alias the image as depicted in Figure 8-1 (page 199). There is always a loss of resolution when changing to a lower resolution format (digital decimation).

8.5. REFERENCES

1. S. K. Park and R. A. Schowengerdt, "Image Sampling, Reconstruction and the Effect of Sample-scene Phasing," *Applied Optics*, Vol. 21(17), pp. 3142-3151 (1982).
2. S. E. Reichenbach, S. K. Park, and R. Narayanswamy, "Characterizing Digital Image Acquisition Devices," *Optical Engineering*, Vol. 30(2), pp. 170-177 (1991).
3. W. Wittenstein, J. C. Fontanella, A. R. Newberry, and J. Baars, "The Definition of the OTF and the Measurement of Aliasing for Sampled Imaging Systems," *Optica Acta*, Vol. 29(1), pp. 41-50 (1982).
4. J. C. Felz, "Development of the Modulation Transfer Function and Contrast Transfer Function for Discrete Systems, Particularly Charge Coupled Devices," *Optical Engineering*, Vol. 29(8), pp. 893-904 (1990).
5. S. K. Park, R. A. Schowengerdt, and M. Kaczynski, "Modulation Transfer Function Analysis for Sampled Image Systems," *Applied Optics*, Vol. 23(15), pp. 2572-2582 (1984).
6. L. deLuca and G. Cardone, "Modulation Transfer Function Cascade Model for a Sampled IR Imaging System," *Applied Optics*, Vol. 30(13), pp. 1659-1664 (1991).
7. J. E. Greivenkamp, "Color Dependent Optical Prefilter for the Suppression of Aliasing Artifacts," *Applied Optics*, Vol. 29(5), pp 676-684 (1990).
8. T. S. Lomheim, L. W. Schumann, R. M. Shima, J. S. Thompson, and W. F. Woodward, "Electro-Optical Hardware Considerations in Measuring the Imaging Capability of Scanned Time-delay-and-integrate Charge-coupled Imagers," *Optical Engineering*, Vol. 29(8), pp. 911-927 (1990).
9. C. M. Webb, "MRTD, How Far Can We Stretch It?" in *Infrared Imaging Systems: Design, Analysis, Modeling, and Testing V*, G. C. Holst, ed., SPIE Proceedings Vol. 1689, pp. 297-307 (1994).

9

MTF THEORY

The optical transfer function (OTF) plays a key role in the theoretical evaluation and optimization of an optical system. The modulation transfer function (MTF) is the magnitude and the phase transfer function (PTF) is the phase of the complex-valued OTF. When an ideal system is viewing incoherent illumination, the OTF is real-valued and positive so that the OTF and MTF are equal. When focus errors or aberrations are present, the OTF may become negative or even complex valued. Electronic circuitry also can be described by an MTF and PTF. The combination of the optical MTF and the electronic MTF creates the electro-optical imaging system MTF. The MTF is the primary parameter used for system design, analysis and specifications.

Since the MTF is the magnitude of the OTF, it is represented by positive values. Phase reversals are represented by a 180-degree shift in the PTF. Nearly all MTFs used for imaging systems are well behaved in the sense that the PTF can be zero. Optical systems are designed with minimal aberration and usually the PTF is sufficiently small that it can be neglected. The notable exception is the detector MTF and linear motion. These MTFs are $\sin(x)/x$ functions whose phase changes at $x = \pi, 2\pi, \ldots$ In this text, rather than use two graphs (MTF and PTF), the phase shifts are represented as negative values on the MTF curve.

When coupled with the three-dimensional noise parameters, the MTF and PTF uniquely define system performance. The MTF and PTF are measures of how the system responds to spatial frequencies. They do not contain any signal intensity information.

The MTF is a measure of how well the system will faithfully reproduce the scene. The highest spatial frequency that can be faithfully reproduced is the system cutoff frequency. Systems can *detect* signals whose spatial frequencies are above cutoff but cannot faithfully reproduce them. For example, a very high-frequency bar pattern may appear as one low contrast blob. From a design point of view, the MTF should be "high" over the spatial frequencies of interest; this range of frequencies is application-specific.

The system MTF and PTF alter the image as it passes through the circuitry. For linear-phase-shift systems, the PTF is of no special interest since it only indicates a spatial or temporal shift with respect to an arbitrarily selected origin.

An image where the MTF is drastically altered is still recognizable whereas large nonlinearities in the PTF can destroy recognizability. Modest PTF nonlinearity may not be noticed visually except those applications where target geometric properties must be preserved (i.e., mapping or photogrammetry). Generally, PTF nonlinearity increases as the spatial frequency increases. Since the MTF is small at high spatial frequencies, the nonlinear-phase-shift effect is diminished.

The MTF is the ratio of output modulation to input modulation normalized to unity at zero spatial frequency. While the modulation changes with system gain, the MTF does not. The input can be as small as desired (assuming a noiseless system with high gain). Or it can be as large as desired since the system is assumed not to saturate.

Four conditions must be met to use MTF theory: (1) the radiation is incoherent, (2) the signal processing is linear, (3) the image is spatially invariant, and (4) the system mapping is single-valued (non-noisy and not digitized). These last three conditions are violated by most electro-optical imaging systems. The optical system will be spatially variant if its impulse response varies from the center to the edge of the field-of-view due to aberrations. Each detector is noisy and this violates the one-to-one mapping requirement. The analog electronics may be noisy and nonlinear image processing may also be present.

A linear system merely modifies the amplitude and phase of the target. No new frequencies are generated in the process. Sampling can create new frequencies. The detector array samples the scene in two directions. These effects were discussed in Chapter 8, *Sampling Theory*, page 199.

No system is truly linear. They may be considered "globally" shift invariant on a macro-scale. As a target moves from the top of the field-of-view to the bottom, the image also moves from the top to the bottom. On a micro-scale, moving a point source across a single detector does not change the detector output. That is, the system is not linear on a micro-scale. In spite of the disclaimers mentioned, electro-optical systems are treated as quasi-linear over a restricted operating region to take advantage of the wealth of mathematical tools available to analyze linear systems.

For modeling purposes, electro-optical systems are characterized as linear spatial-temporal systems that are shift-invariant with respect to both time and two spatial dimensions. Although space is three-dimensional, an imaging system displays only two dimensions.

Time and spatial coordinates are treated separately. For example, optical elements do not generally change with time and therefore are characterized only by spatial coordinates. Similarly, electronic circuitry exhibits only temporal responses. The detector provides the interface between the spatial and temporal components and its response depends on both temporal and spatial quantities. The conversion of two-dimensional optical information to a one-dimensional electrical response assumes a linear photo-detection process. Implicit in the detector MTF is the conversion from input flux to output voltage.

Time filters are different from spatial filters in two ways. Time filters are single-sided in time and must satisfy the causality requirement that no change in the output may occur before the application of an input. Optical filters are double-sided in space. Electrical signals may be either positive or negative whereas optical intensities are always positive. As a result, optical designers and circuit designers use different terminology.

9.1. MTF DEFINITION

Modulation is the variation of a sinusoidal signal about its average value (Figure 9-1). It can be considered as the AC amplitude divided by the DC level. The modulation is:

$$MODULATION = M = \frac{V_{MAX} - V_{MIN}}{V_{MAX} + V_{MIN}} = \frac{AC}{DC} \qquad (9\text{-}1)$$

V_{MAX} and V_{MIN} are the maximum and minimum signal levels respectively. The modulation transfer function is the output modulation produced by the system divided by the input modulation at that frequency:

$$MTF = \frac{OUTPUT\ MODULATION}{INPUT\ MODULATION} \qquad (9\text{-}2)$$

The concept is presented in Figure 9-2. Three input and output signals are plotted in Figures 9-2a and 9-2b and the resultant MTF is shown in Figure 9-2c. As a ratio, the MTF is a relative measure whose value range from zero to one.

222 CCD ARRAYS, CAMERAS, and DISPLAYS

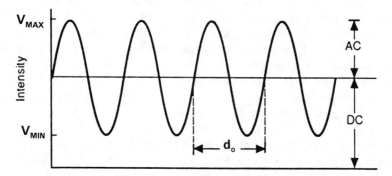

Figure 9-1. Definition of Target Modulation. d_o is the extent of one cycle. If d_o is measured at the detector (in millimeters), then the spatial frequency is $f_o = 1/d_o$ cy/mm.

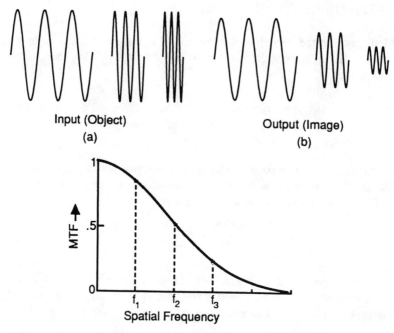

Figure 9-2. Modulation transfer function. (a) Input signal for three different spatial frequencies, (b) output for the three frequencies, and (c) the MTF is the ratio of output-to-input modulation.

MTF THEORY 223

9.2. LINEAR FILTER THEORY

Linear filter theory was developed for electronic circuitry and has been extended to optical, electro-optical, and mechanical systems. Linear filter theory forms an indispensable part of system analysis.

Let $S\{\ \}$ be a linear operator that maps one function, $f(x)$, into another function, $g(x)$:

$$S\{f(x)\} = g(x) \qquad (9\text{-}3)$$

Let the response to two inputs, $f_1(x)$ and $f_2(x)$, be $g_1(x)$ and $g_2(x)$:

$$S\{f_1(x)\} = g_1(x) \quad and \quad S\{f_2(x)\} = g_2(x) \qquad (9\text{-}4)$$

For a linear system, the response to a sum of inputs is equal to the sum of responses to each input acting separately. For any arbitrary scale factors, the superposition principle states

$$S\{a_1 f_1(x) + a_2 f_2(x)\} = a_1 g_1(x) + a_2 g_2(x) \qquad (9\text{-}5)$$

9.2.1. The EO SYSTEM as a LINEAR SYSTEM

An object can be thought of as the sum of an infinite array of impulses (Dirac delta functions) located inside the target boundaries. Thus, the object can be decomposed into a series of weighted Dirac delta functions, $\delta(x-x_o)$, $\delta(y-y_o)$:

$$O(x,y) = \sum_{x_o=-\infty}^{\infty} \sum_{y_o=-\infty}^{\infty} O(x_o, y_o) \delta(x - x_o) \delta(y - y_o) \Delta x_o \Delta y_o \qquad (9\text{-}6)$$

As Δx_o and Δy_o decrease in size, Equation 9-6 becomes an integral. Figure 9-3 illustrates one-dimensional decomposition. Similarly, an electronic wave form in time can be decomposed into a series of weighted Dirac delta functions.

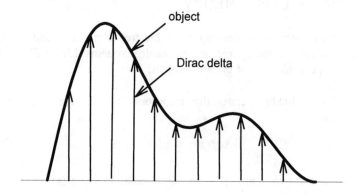

Figure 9-3. An electronic wave form or object can be decomposed into a series of closely spaced impulses whose amplitude is equal to the wave form at that value.

Using the superposition principle, $S\{\ \}$ operates on each individual input. Mathematically,

$$(9\text{-}7) \quad I(x,y) = \sum_{x_0=-\infty}^{\infty} \sum_{y_0=-\infty}^{\infty} S\{O(x_o,y_o)\delta(x-x_o)\delta(y-y_o)\Delta x_o \Delta y_o\}$$

The object $O(x_o, y_o)$ can be considered the weightings, a_i, in Equation 9-5. That is, the input has been separated into a series of functions $a_1 f_1(x) + a_2 f_2(x) + \ldots$ For small increments, this becomes the convolution integral

$$I(x,y) = \int_{-\infty}^{\infty} \int_{-\infty}^{\infty} O(x_o,y_o)\, S\{\,\delta(x-x_o)\delta(y-y_o)\}\, dx_o\, dy_o \qquad (9\text{-}8)$$

and is symbolically represented by

$$I(x,y) = O(x,y) * S(x,y) \qquad (9\text{-}9)$$

where $*$ indicates the convolution operator. $S\{\ \}$ is the system's response to an input impulse. The impulse creates the point spread function for the optical systems and impulse response for the electronic circuitry. The individual outputs are added to produce the image (Figure 9-4).

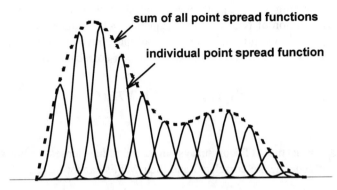

Figure 9-4. The linear operator, S{ }, transforms each input Dirac delta into an output point spread function. The sum of the point spread functions creates the image.

If the image is passed through another linear system, the superposition principle is applied again:

$$S_2\{g(x)\} = S_2\{S_1\{f(x)\}\ \} \qquad (9\text{-}10)$$

or

$$I(x,y) = O(x,y) * S_1(x,y) * S_2(x,y) \qquad (9\text{-}11)$$

9.2.2. CASCADING MTFs

Taking the Fourier transform of Equation 9-9 shows that the convolution operator becomes a multiplication in the frequency domain. When MTFs are multiplied together, they are said to be cascaded. The magnitude of Fourier transform of the impulse response is the MTF of the circuit. The two-dimensional MTF is the magnitude of the complex-valued two-dimensional Fourier transform of the point spread function (PSF). The PSF is the system response when viewing an ideal point source.

The convolution theorem of Fourier analysis states that the Fourier transform of the convolution of two functions (Equation 9-11) equals the product of the transform of the two functions. That is, multiple convolutions are equivalent to multiplication in the frequency domain. For non-correlated MTFs,

$$MTF_{SYS}(f) = \prod_{i=1}^{N} MTF_i(f) \qquad (9\text{-}12)$$

where MTF_i is the MTF of subsystem i. Individual lenses within an optical system, in general, are correlated and their MTFs cannot be cascaded.

The two-dimensional system MTF is

$$MTF_{SYS}(f_x, f_y) = MTF_{OPTICS}(f_x, f_y)\, MTF_{DETECTOR}(f_x, f_y)\, MTF_{ELEC}(f_{Hz}) \qquad (9\text{-}13)$$

f_x and f_y are the spatial frequencies in the horizontal and vertical directions respectively. The electrical frequency, f_{Hz}, must be appropriately scaled into spatial frequency. For convenience, the MTF is assumed separable in two orthogonal axes (usually coincident with the detector array axes) to obtain two one-dimensional MTFs:

$$MTF_{SYS}(f_x, f_y) \approx MTF_{SYS}(f_x)\, MTF_{SYS}(f_y) \qquad (9\text{-}14)$$

Electrical filters are causal, one-dimensional, and are considered to operate in the direction of the readout clocking only. Thus, the horizontal and vertical MTFs may be different. The one-dimensional MTF is the magnitude of the Fourier transform of the line spread function, LSF. The LSF is the resultant image produced by the imaging system when it is viewing an ideal line.

9.3. SUPERPOSITION APPLIED TO OPTICAL SYSTEMS

If the MTF of a system is known, one can compute the image for any arbitrary object. First, the object is dissected into its constituent spatial frequencies (i.e., the Fourier transform of the object is obtained). Next, each of these frequencies is multiplied by the system MTF at that frequency. Then the inverse Fourier transform provides the image.

MTF THEORY 227

To illustrate the superposition principle and MTF approach, we will show how an ideal optical system modifies an image. An ideal optical system is, by definition, a linear-phase-shift system. The most popular test target consists of a series of bars - typically three or four bars although more may be used. For illustrative purposes, the periodic bars are assumed to be of infinite extent. A one-dimensional square wave, when expanded into a Fourier series about the origin, contains only odd harmonics:

$$f(x) = \frac{4}{\pi} \sum \frac{1}{n} \sin\left(\frac{2\pi n x}{d_o}\right) \quad n = 1, 3, 5, \ldots \qquad (9\text{-}15)$$

where d_o is the period of the square wave. The fundamental frequency f_o is $1/d_o$ (see Figure 9-1, page 222). Taking the Fourier transform of the square wave provides discrete spatial frequencies $1/d_o$, $3/d_o$, $5/d_o$, ... whose amplitudes are $4/\pi$, $4/3\pi$, $4/5\pi$, ... respectively.

Let a circular optical system image the square wave. The MTF for a circular, clear aperture, diffraction-limited lens is

$$MTF_{DIFF} = \frac{2}{\pi}\left[\cos^{-1}\left(\frac{f_x}{f_{OC}}\right) - \frac{f_x}{f_{OC}}\sqrt{1 - \left(\frac{f_x}{f_{OC}}\right)^2}\right] \qquad (9\text{-}16)$$

when $f_x < f_{OC}$. The optical cutoff in image space is $f_{OC} = D_O/(\lambda\, fl)$ where D_O is the aperture diameter, λ is the average wavelength, and fl is the focal length.

By superposition, the optical MTF and square wave amplitudes are multiplied together at each spatial frequency:

$$I(f_x) = MTF_{DIFF}(f_x)\, O(f_x) \qquad (9\text{-}17)$$

Taking the inverse Fourier transform provides the resultant image. Equivalently,

$$I(x) = MTF_{DIFF}(f_o)\left[\frac{4}{\pi}\sin(2\pi x f_o)\right]$$
$$+ MTF_{DIFF}(3f_o)\left[\frac{4}{3\pi}\sin(6\pi x f_o)\right] + \ldots \qquad (9\text{-}18)$$

If f_o is greater than $f_{OC}/3$, only the fundamental of the square wave will be faithfully imaged by the optical system. Here, the square wave will appear as a sine wave. Note that the optical MTF will reduce the image amplitude. As f_o decreases, the image will look more like a square wave (Figure 9-5). Since the optics does not pass any frequencies above f_{OC}, the resultant wave form is a truncation of the original series modified by the optical MTF. This results in some slight ringing. This ringing is a residual effect of the Gibbs phenomenon. Square wave analysis is further considered in the following section, *Contrast Transfer Function*.

9.4. CONTRAST TRANSFER FUNCTION

The system response to a square-wave target is the contrast transfer function (CTF) or square-wave response (SWR). The CTF is a convenient measure because of the availability of square wave targets. It is not a transfer function in the same sense that the MTF is and, as such, subsystem CTFs cannot be cascaded. Figure 9-6 illustrates the relationship between square wave and sinusoidal amplitudes. The CTF is typically higher than the MTF at all spatial frequencies.

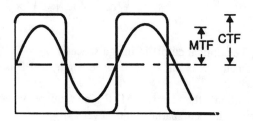

Figure 9-6. AC components of the CTF and MTF. The CTF is usually equal to or greater than the MTF.

MTF THEORY 229

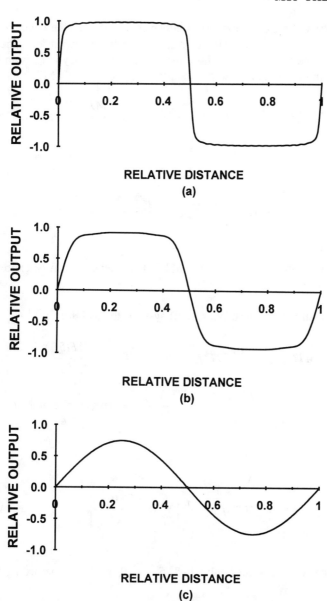

Figure 9-5. Square waves imaged by a circular aperture. As the square wave fundamental frequency increases, the edges become rounded. Eventually, the square wave will appear as a sinusoid when $f_o \geq f_{OC}/3$. (a) $f_o = f_{OC}/40$, (b) $f_o = f_{OC}/10$, and (c) $f_o = f_{OC}/3$.

230 CCD ARRAYS, CAMERAS, and DISPLAYS

A square wave can be expressed either as an infinite sine (Equation 9-15, page 227) or cosine series. Coltman[1] selected the cosine expansion. The output amplitude of the square wave, whose fundamental frequency is f_o, is an infinite sum of the input cosine amplitudes modified by the system's MTF:

$$CTF(f_o) = \frac{4}{\pi} \left| MTF(f_o) - \frac{MTF(3f_o)}{3} + \frac{MTF(5f_o)}{5} - \ldots \right| \quad (9\text{-}19)$$

or

$$CTF(f_o) = \frac{4}{\pi} \left| \sum_{k=0}^{\infty} (-1)^k \frac{MTF[(2k+1)f_o]}{2k+1} \right| \quad (9\text{-}20)$$

By adding and subtracting terms, the MTF at frequency f_o can be expressed as an infinite sum of CTFs.

Again, assuming a linear-phase-shift system, the MTF is[1,2]:

$$MTF(f_o) = \frac{\pi}{4} \left| CTF(f_o) + \frac{CTF(3f_o)}{3} - \frac{CTF(5f_o)}{5} + \frac{CTF(7f_o)}{7} + \frac{CTF(11f_o)}{11} + irregular\ terms \right| \quad (9\text{-}21)$$

or

$$MTF(f_x) = \frac{\pi}{4} \left| \sum_{k=0}^{\infty} B_k \frac{CTF(kf_x)}{k} \right| \quad (9\text{-}22)$$

where k takes on odd values only: 1, 3, 5, ... B_k is -1 or 1 according to:

$$B_k = (-1)^m (-1)^{\frac{k-1}{2}} \quad for\ r = m \quad (9\text{-}23)$$

r is the number of different prime factors in k. m is the total number of primes into which k can be factored. $B_k = 0$ for $r < m$.

MTF THEORY

Theoretically, to obtain the MTF at frequency f_o, an infinite number of square wave responses must be measured. However, the number required is limited by the spatial cutoff frequency, $f_{MTF=0}$, where the MTF approaches zero and remains zero there after. For bar targets whose spatial frequency is above ⅓ $f_{MTF=0}$, the MTF is equal to $\pi/4$ times the measured CTF. That is, above ⅓ $f_{MTF=0}$, only one target is necessary to compute[3] the MTF. This was illustrated in Section 9.3., *Superposition Applied to Optical Systems*, page 226.

For low spatial frequencies, the output nearly replicates the input (Figure 9-7a). For frequencies near ⅓ cutoff, the output starts to appear sinusoidal (Figure 9-7b). Above ⅓ cutoff (Figure 9-7c), the MTF amplitude is $\pi/4$ times the CTF amplitude.

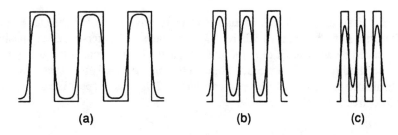

Figure 9-7. Input and output wave forms for a linear system. (a) Very low frequency signals are faithfully reproduced, (b) mid-spatial frequencies tend to look like sinusoids, and (c) input square waves whose spatial frequencies are above ⅓ $f_{MTF=0}$ appear as sinusoidal outputs.

The relationship between the CTF and MTF was developed for analog optical systems. Is appropriate for wet-film photographic cameras and image intensifiers. It can also be applied in the scan direction for analog scanning devices (vidicons and displays). For displays, an important resolution measure is based on the visibility of square waves: the TV limiting resolution.

For sampled data systems, the mathematical relationship between CTF and MTF has not yet been developed. Therefore the relationship should be used cautiously for sampled data systems, if at all. CTF measurements are appropriate tests for system performance verification. It is the conversion to MTF that has not been validated. The CTF and TV limiting resolution are inappropriate metrics for flat panel displays. As flat panel display technology matures, new resolution metrics will evolve.

232 CCD ARRAYS, CAMERAS, and DISPLAYS

Sampling is present in all CCD imaging systems due to the discrete location of the detectors. Phasing effects between the detector and the location of the target introduce problems at nearly all spatial frequencies. The interaction of the bar frequency and the system sampling frequency produces sum and difference frequencies that produce variations in amplitude and bar width (see in Section 8.3.3., *Image Distortion*, page 212).

The CTF is well behaved, in the sense that the bar width remains constant, when the bar spatial frequency is proportional to the system Nyquist frequency:

$$f_o = \frac{f_N}{k} \qquad (9\text{-}24)$$

where k is an integer. These spatial frequencies provide the in-phase CTF. For CCD arrays, $f_N = fl/(2\,d_{CC})$ where d_{CC} is the detector center-to-center spacing (also called detector pitch or pixel pitch). f_N is the highest spatial frequency that can be faithfully reproduced. The CTF can be used to assess system performance at the spatial frequencies selected. It is usually specified only at the Nyquist frequency (k = 1).

9.5. REFERENCES

1. J. W. Coltman, "The Specification of Imaging Properties by Response to a Sine Wave Input," *Journal of the Optical Society of America*, Vol. 44(6), pp. 468-471 (1954).
2. I. Limansky, "A New Resolution Chart for Imaging Systems," *The Electronic Engineer*, Vol. 27(6), pp. 50-55 (1968).
3. G. C. Holst, *Testing and Evaluation of Infrared Imaging Systems*, pp. 238-246, JCD Publishing, Winter Park, FL (1993).

10

SYSTEM MTF

The equations describing array (Chapter 4) and camera (Chapter 6) performance assume that the image is very large and it illuminates many detector elements. That is, the object contained only very low spatial frequencies and that the system modulation transfer function (MTF) is essentially unity over these spatial frequencies.

The optical and detector MTFs modify the signal spatial frequencies. These frequencies may also be affected by electronic filter MTFs although these electronic filter functions are typically designed to pass the entire image frequency spectrum without attenuation.

Noise, on the other hand, originates at the detector and within the electronic circuitry. It is modified by the electronic MTFs only. Detector array noise is fixed but analog amplifier noise increases as the electronic bandwidth increases. The camera SNR is maximized when the electronic bandwidth passes the signal with minimal degradation. Wider bandwidth electronics will increase the noise but not the signal and thereby reduce the SNR.

For mathematical convenience, a specific target feature can be associated with a fictitious spatial frequency (Figure 10-1). The SNR for this feature is

$$SNR(f_o) = MTF(f_o)\frac{n_{pe}}{\langle n_{SYS} \rangle} \quad (10\text{-}1)$$

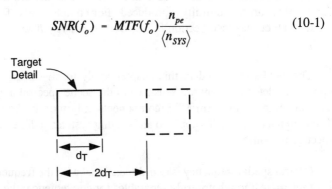

Figure 10-1. Fictitious spatial frequency associated with a target feature. If the image detail size is d_T mm at the detector plane, then the fictitious frequency is $f_o = 1/(2d_T)$ cy/mm.

234 CCD ARRAYS, CAMERAS, and DISPLAYS

$\langle n_{SYS}\rangle$ includes shot, reset, amplifier, quantization, and fixed pattern noise sources (see Section 4.2., *Noise*, page 104). The total amplifier noise depends upon the electronic bandwidth and that depends on the electronic frequency response associated with the readout signal chain.

A CCD camera consists of optics, an array of pixels, and a signal chain with electronic frequency response characteristics. The complete imaging system includes the display:

$$MTF_{SYS} = MTF_{OPTICS}MTF_{DETECTOR}MTF_{SIGNAL-CHAIN}MTF_{DISPLAY} \quad (10\text{-}2)$$

The observer processes this information to create a perceived MTF:

$$MTF_{PERCEIVED} = MTF_{SYS}MTF_{EYE} \quad (10\text{-}3)$$

When $MTF_{PERCEIVED}$ is coupled to system noise, system responsivity, and the eye integration capabilities, it is possible to predict the minimum resolvable contrast (described in Chapter 12, *Minimum Resolvable Contrast*). MRC is used by the military to predict target detection and recognition ranges for a given probability of success.

Since CCD arrays are undersampled, the highest frequency that can be reproduced is the array Nyquist frequency, f_N. As a result, it is common practice to evaluate Equation 10-1 at $f_o = f_N$. This dictates the smallest target feature that can be faithfully reproduced. For completeness, f_N should be added to all MTF curves (see Section 8.3.2, *Detector Array Nyquist Frequency*, page 209).

The MTFs presented in this chapter apply to all staring arrays. Only the optics and detector array are sensitive to the scene spectral and spatial content. The remaining subsystem MTF components only modify the electrical signals. These signal chain MTFs apply to all imaging systems independent of the system spectral response.

Optical spatial frequency is two-dimensional with the frequency ranging from $-\infty$ to $+\infty$. Although not truly separable, for convenience, the horizontal MTF, $MTF(f_x)$, and vertical MTF, $MTF(f_y)$, are analyzed separately. Electrical filters are causal, one-dimensional, and are considered to operate in the direction of the readout clocking only. Thus, the horizontal and vertical MTFs may be different.

SYSTEM MTF 235

Complete characterization requires both the modulation and the phase transfer functions (PTF). For linear-shift-invariant systems, the PTF only indicates a shift from an arbitrarily selected origin. Phase shifts of 180° can occur within detectors and when linear motion is present [sinc(x) functions]. In this text, rather than use two graphs (MTF and PTF), the phase shifts are represented as negative values on the MTF curve.

The symbols used in this book are summarized in the *Symbol List* (page xiv) and it appears after the *Table of Contents*.

10.1. SPATIAL FREQUENCY

There are four different locations where spatial frequency domain analysis is appropriate. They are object space (before the camera optics), image space (after the camera optics), monitor screen, and observer (at the eye) spatial frequencies. Simple equations relate the spatial frequencies in all these domains.

Figure 10-2 illustrates the spatial frequency associated with a bar target. Bar patterns are the most common test targets and are characterized by their fundamental spatial frequency. Using the small angle approximation, the angle subtended by one cycle (one bar and one space) is d_o/R_1 where d_o is the spatial extent of one cycle and R_1 is the distance from the imaging system entrance aperture to the target. When using a collimator to project the targets at apparently long ranges, the collimator focal length replaces R_1 so that targets placed in the collimator's focal plane can be described in object space. The horizontal object-space spatial frequency, f_{OS}, is the inverse of the horizontal target angular subtense and is usually expressed in cycles/mrad:

$$f_{OS} = \frac{1}{1000}\left(\frac{R_1}{d}\right) \quad \frac{cycles}{mrad} \qquad (10\text{-}4)$$

The object space domain is used by the military for describing system performance (discussed in Chapter 12, *Minimum Resolvable Contrast*)

Optical designers typically quote spatial frequencies in image space to specify the resolving capability of lens systems. Photographic cameras and CCD cameras are typically specified with spatial frequencies in image space and are used in this chapter.

236 CCD ARRAYS, CAMERAS, and DISPLAYS

It is the object-space spatial frequency divided by the system focal length:

$$f_i = \frac{f_{OS}}{fl} \quad \frac{\text{line-pairs}}{mm} \quad \text{or} \quad \frac{\text{cycles}}{mm} \quad (10\text{-}5)$$

f_i is the inverse of one cycle in the focal plane of the lens system. Although used interchangeably, line-pairs suggest square wave targets and cycles suggest sinusoidal targets. To maintain dimensionality, if f_{OS} is measured in cy/mrad then the focal length must be measured in meters to obtain cy/mm.

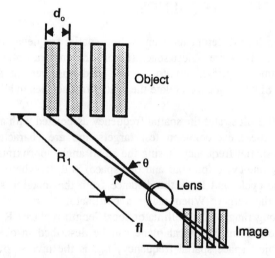

Figure 10-2. Correspondence of spatial frequencies in object and image space. Although the MTF is defined for sinusoidal signals, the bar target is the most popular test target (shown here). Bar targets are characterized by their fundamental frequency.

The resolving capability of a monitor can be specified by the maximum number of vertical lines seen (Figure 10-3). By convention, although the bar width, Δx, is measured in the horizontal direction, the number of lines is expressed relative to the picture height, PH. The number of lines may be specified as TV lines per picture height, lines per picture height or simply lines:

$$N_m = \frac{PH}{\Delta x} \quad (10\text{-}6)$$

There are two TV lines per cycle. Therefore, the spatial frequency at the monitor screen is

$$f_m = \frac{N_m}{2} = \frac{PH}{2\Delta x} \quad \frac{cycles}{picture\ height} \qquad (10\text{-}7)$$

The size of the image can be measured with a ruler. Using the monitor aspect ratio, α_m, f_m can be converted to a spatial frequency normalized to the monitor width:

$$f_{mx} = \alpha_m f_m \quad \frac{cycles}{picture\ width} \qquad (10\text{-}8)$$

For some computer monitors, $\alpha_m = 1$. For commercial televisions, $\alpha_m = 4/3$ and for HDTVs, $\alpha_m = 16/9$.

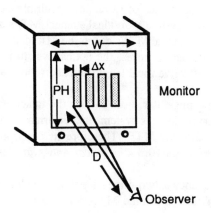

Figure 10-3. Correspondence of spatial frequencies at the monitor screen and observer's eye.

The spatial frequency presented to the observer depends upon the observer's viewing distance and the image size on the monitor. In the visual psychophysical literature[1], the usual units are cycles/deg:

$$f_{EYE} = \frac{1}{2\tan^{-1}\left(\frac{\Delta x}{D}\right)} \quad \frac{cycles}{deg} \quad (10\text{-}9)$$

Here, the arc tangent is expressed in degrees. The small angle approximation may not be valid for the observer spatial frequency since the angle subtended can become quite large as the observer moves toward the monitor.

10.2. OPTICS MTF

Optical systems consist of several lenses or mirror elements with varying focal lengths and varying indices of refraction. Multiple elements are used to minimize lens aberrations. While individual element MTFs are used for design and fabrication, these individual MTFs cannot be cascaded to obtain the optical system MTF. For modeling purposes, the optical system is treated as a single lens that has the same effective focal length and aberrations as the lens system.

The optical anti-aliasing filter, used to blur the image on single chip color filter array, is also part of the lens system. Its performance is best described by the way it affects the detector MTF. Therefore its description is included the detector section (i.e., Section 10.3.4., *Optical Anti-Alias Filter*).

Optical spatial frequency is two-dimensional with the frequency ranging from $-\infty$ to $+\infty$. The diffraction-limited MTF for a circular aperture (in image space) is

$$MTF_{OPTICS} = \frac{2}{\pi}\left[\cos^{-1}\left(\frac{f_i}{f_{OC}}\right) - \frac{f_i}{f_{OC}}\sqrt{1-\left(\frac{f_i}{f_{OC}}\right)^2}\right] \quad (10\text{-}10)$$

when $f_i < f_{OC}$ and is zero elsewhere.

The optical cutoff is $f_{OC} = D_O/(\lambda \, fl) = 1/(F\lambda)$. D_O is the aperture diameter and fl is the effective focal length. The f-number, F, is equal to fl/D_O. Diffraction limited optics exhibit radial symmetry and $f_i^2 = f_x^2 + f_y^2$. Although not truly separable, for convenience only $MTF(f_x)$ and $MTF(f_y)$ are calculated. For an aberration free and radially symmetric optical system, MTF_{OPTICS} is the same in the horizontal direction and vertical directions.

Figure 10-4 illustrates MTF_{OPTICS} as a function of f_i/f_{OC}. This equation is only valid for monochromatic light and the cutoff frequency is dependent upon the wavelength. The extension to polychromatic light is lens specific. Most lens systems are color corrected (achromatized) and therefore there is no simple way to apply this simple formula to predict the MTF. As an approximation to the polychromatic MTF, the average wavelength, is used to calculate the cutoff frequency:

$$\lambda_{ave} \approx \frac{\lambda_{MAX} + \lambda_{MIN}}{2} \qquad (10\text{-}11)$$

For example, if the system spectral response ranges from 0.4 to 0.7 μm, then $\lambda_{AVE} = 0.55 \, \mu$m.

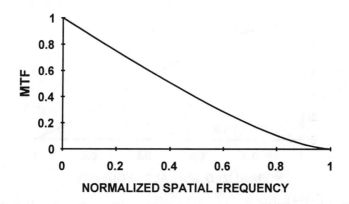

Figure 10-4. MTF_{OPTICS} for a circular aperture normalized to f_i/f_{OC}.

For most systems operating in the visible spectral region, the optical system may be considered near diffraction limited and in focus. Approximations for aberrated and defocused optical systems may be found in reference 2. The MTFs for optical systems with rectangular apertures[3] or telescopes with a central obscuration (Cassegranian optics) are also available[4].

10.3. DETECTORS

The two-dimensional spatial response of a rectangular detector is

$$MTF_{DETECTOR}(f_x, f_y) = sinc(d_H f_x) sinc(d_V f_y) \quad (10\text{-}12)$$

where $sinc(x) = sin(\pi x)/(\pi x)$. d_H and d_V are the photosensitive detector sizes in the horizontal and vertical directions respectively. Note that arrays are often specified by the pixels size. With 100% fill-factor arrays, the pixel size is equal to the detector size. With finite fill-factor arrays, the photosensitive detector dimensions are less than the pixel dimensions. The effective optical size may be modified optically by microlenses or an anti-alias filter.

Although the detector MTF is valid for all spatial frequencies from $-\infty$ to $+\infty$, it is typically plotted up to the first zero (called the cutoff) which is occurs at $f_i = 1/d$. Figure 10-5 illustrates the full MTF without regard to the array Nyquist frequency. The analyst must add the array Nyquist frequency to the curve (see Section 8.3.2, *Detector Array Nyquist Frequency*, page 209). The MTF of virtual phase detectors is a complex expression[5].

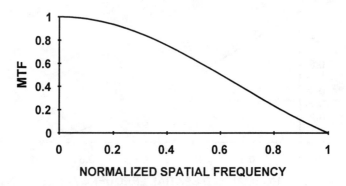

Figure 10-5. Typical detector MTF representation as a function of normalized spatial frequency df_i. The MTF is usually plotted up to the first zero. The array Nyquist frequency must be added. $f_N = 1/(2d_{CC})$

10.3.1. DIFFUSION MTF

As the wavelength increases, photon absorption occurs at increasing depths in the detector material (see Section 4.1.1., *Spectral Response*, page 92). A photoelectron generated deep within the substrate will experience a three-dimensional random walk until it recombines or reaches the edge of a depletion region where the pixel electric field exists. A photoelectron generated under one well may eventually land in an adjoining well. This blurs the image and its effect is described by an MTF. Beyond 0.8 μm, diffusion (random walk) affects the detector MTF and the response is expressed by $MTF_{DETECTOR}MTF_{DIFFUSION}$. Imagery produced by near-IR sources will appear slightly blurry compared to imagery produced by a visible source.

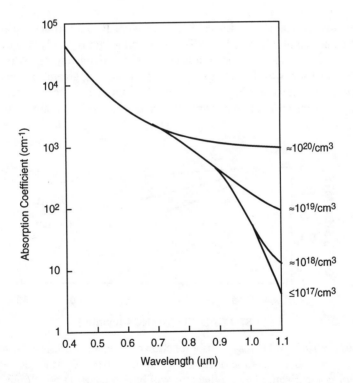

Figure 10-6. Silicon absorption coefficient as a function of wavelength and doping concentration at 25° C. Detectors typically have doping concentrations less than $10^{17}/cm^3$. From reference 7.

The diffusion MTF[6] for a front illuminated device is

$$MTF_{DIFFUSION}(f_i) = \frac{1 - \dfrac{\exp(-\alpha_{ABS}L_D)}{1 + \alpha_{ABS}L(f)}}{1 - \dfrac{\exp(-\alpha_{ABS}L_D)}{1 + \alpha_{ABS}L_{DIFF}}} \qquad (10\text{-}13)$$

L_D is the depletion width and $L(f)$ is the frequency-dependent component of the diffusion length given by

$$L(f_i) = \frac{L_{DIFF}}{\sqrt{1 + (2\pi L_{DIFF} f_i)^2}} \qquad (10\text{-}14)$$

α_{ABS} is the spectral absorption coefficient. Figure 10-6 illustrates the absorption coefficient in bulk p-type silicon. The diffusion length typically ranges between 50 μm and 200 μm. The depletion width is approximately equal to the gate width. For short wavelengths ($\lambda < 0.6$ μm), α_{ABS} is large and lateral diffusion is negligible. Here, $MTF_{DIFFUSION}$ approaches unity. For the near IR ($\lambda > 0.8$ μm), the diffusion MTF may dominate the detector MTF (Figure 10-7).

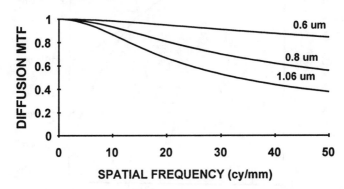

Figure 10-7. $MTF_{DIFFUSION}$ as a function of wavelength. $L_D = 10$ μm and $L_{DIFF} = 100$ μm. As the diffusion length increases, the MTF decreases. For comparison, a 20 μm detector will have a cutoff at 50 cy/mm.

To account for the effects of lateral diffusion, Schmalz and Lomheim[8] introduced a trapezoidal response approximation. This MTF only applies in the direction perpendicular to charge transport:

$$MTF_{DIFFUSION}(f_i) \approx sinc(d_{CC}f_i)\, sinc[(d_{CC}-\beta)f_i] \quad (10\text{-}15)$$

d_{CC} is the detector pitch and β is the region over which the detector responsivity is flat (Figure 10-8). At the boundary between pixels, the trapezoidal response is 0.5 indicating equal probability of an electron going to either pixel. The channel stop region is related to the sloped part of the trapezoidal region.

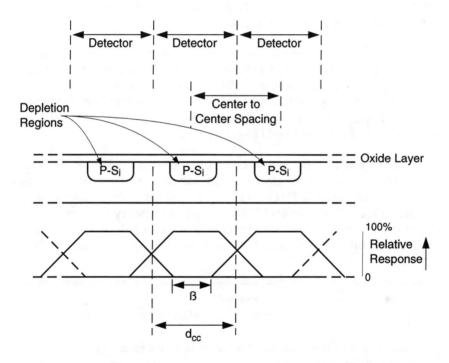

Figure 10-8. Trapezoidal approximation for long-wavelength photon absorption. Most of the charge goes to the charge well nearest to the photon absorption site.

244 CCD ARRAYS, CAMERAS, and DISPLAYS

10.3.2. CHARGE TRANSFER EFFICIENCY

CCD performance is characterized by the transfer efficiency and number of transfers (see section 3.6., *Charge Transfer Efficiency*, page 79). An MTF[9] that accounts for the incomplete transfer of electrons in object space is

$$MTF_{CTE}(f_i) = \exp\left(-N_{TRANS}(1-\epsilon)\left[1-\cos\left(\frac{\pi f_i}{f_N}\right)\right]\right) \qquad (10\text{-}16)$$

N_{TRANS} is the total number of charge transfers from a detector to the output amplifier, and ϵ is the charge transfer efficiency for each transfer. N_{TRANS} is the number of detectors multiplied by the number of CCD phases. If $N_{TRANS}(\epsilon - 1)$ is small, then $MTF(f_N) \approx 1 - 2N_{TRANS}(1-\epsilon)$.

This MTF depends upon the number of charge transfers. If the image is located next to the readout, the number of transfers is small. On the other hand, if the image is at the extreme end of the array, the charge must be transferred across the entire array and through the serial readout register. For an average response, N_{TRANS} should be one-half the maximum number of transfers. This is mathematically equivalent to $(MTF_{CTE})^{\frac{1}{2}}$.

Transfers may be either in the vertical direction (column readout) or in the horizontal direction (row readout). Horizontal readout affects $MTF(f_x)$ and vertical readout affects $MTF(f_y)$. MTF_{CTE} only applies to the transfer readout direction. Figure 10-9 illustrates MTF_{CTE} at Nyquist frequency for several values of transfer efficiency. For consumer devices, CTE > 0.9999 and N_{TRANS} > 1500. For scientific applications, CTE > 0.999999 and N_{TRANS} can exceed 5000.

Example 10-1
Number of Transfers

What is the maximum number of transfers that exist in an array that contains 1000 x 1000 detector elements?

For a charge packet that is furthest from the sense node, the charge must be transferred down one column (1000 pixels) and across 1000 pixels for a total of 2000 pixels. The number of transfers is 4000, 6000, and 8000 for two-phase, three-phase, and four-phase devices respectively.

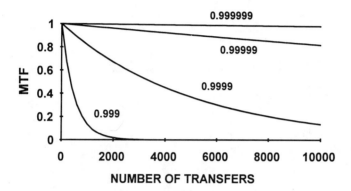

Figure 10-9. MTF$_{CTE}$ at Nyquist frequency for several values of transfer efficiency. As the number of transfers increases, the efficiency must also increase to insure that the MTF is not adversely affected.

10.3.3. TDI

TDI operation was described in Section 3.2.6, *Time-Delay and Integration*, page 66. It is essential that the clock rate match the image velocity so that the photo-generated charge packet is always in sync with the image. Any mismatch will blur the image. Image rotation with respect to the array axes affects both the horizontal (TDI direction) and vertical (readout direction) MTFs simultaneously. For simplicity they are treated as approximately separable - MTF(f_x,f_y) ≈ MTF(f_x)MTF(f_y).

If the TDI direction is horizontal, velocity errors degrades[10] the detector MTF by

$$MTF_{TDI}(f_x) = \frac{\sin(\pi N_{TDI} d_{ERROR} f_x)}{N_{TDI} \sin(\pi d_{ERROR} f_x)} = \frac{sinc(N_{TDI} d_{ERROR} f_x)}{sinc(d_{ERROR} f_x)} \quad (10\text{-}17)$$

where N_{TDI} is the number of TDI elements. d_{ERROR} is the difference between the expected image location and the actual location at the first pixel (Figure 10-10). After N_{TDI} elements, the image is displaced $N_{TDI} d_{ERROR}$ from the desired location. MTF$_{TDI}$ is simply the MTF of an averaging filter for N_{TDI} samples displaced d_{ERROR} one relative to the next. As an averaging filter, the effective sampling rate is $1/d_{ERROR}$.

246 CCD ARRAYS, CAMERAS, and DISPLAYS

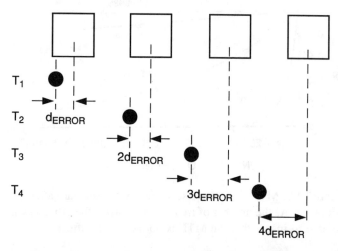

Figure 10-10 Definition of d_{ERROR}. With a velocity mismatch, the target falls further and further away from the desired location (on the center of the detector). After 4 TDI elements, the target is displaced from the detector center by $4d_{ERROR}$.

This is a one-dimensional MTF that apples in the TDI direction only. In image space

$$d_{ERROR} = |\Delta V| t_{INT} \quad mm \qquad (10\text{-}18)$$

ΔV relative velocity error between the image motion and the charge packet "motion" or "velocity." The charge packet "velocity" is d_{CC}/t_{INT} where t_{INT} is the integration time for each pixel.

Figure 10-11 portrays the MTF for two different relative velocity errors for 32 and 96 TDI stages. As the number of TDI stages increases, α_{ERROR} must decrease to maintain the MTF performance. Equivalently, the accuracy of knowing the target velocity must increase as N_{TDI} increases. Since the image is constantly moving, a slight image smear occurs[11] because the photo-generated charge is created at discrete times (e.g., at intervals of t_{CLOCK}). This linear motion MTF (discussed in Section 10.4.1., *Linear Motion*) is often small and therefore is usually neglected.

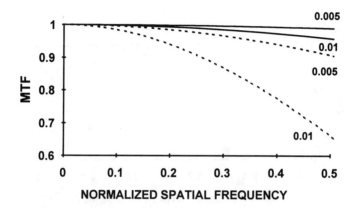

Figure 10-11. MTF degradation due to mismatch between the image velocity and the charge packet "velocity." d_{ERROR}/d_{CC} equals 0.005 and 0.01. The spatial frequency axis is normalized to $d_{CC}f_x$ and the MTF scale is expanded from 0.6 to 1.0. The solid lines represent $N_{TDI} = 32$ and the dashed lines are for $N_{TDI} = 96$.

TDI is also sensitive to image rotation. If the image motion subtends an angle θ with respect to the TDI rows (TDI direction), the image moves vertically $N_{TDI}d_{CCH}\tan(\theta)$ across the array. The MTF degradation in the vertical direction is

$$MTF_{TDI}(f_y) = \frac{\sin\left[\pi\, N_{TDI}\, d_{CCH} \tan(\theta)\, f_y\right]}{N_{TDI}\sin\left[\pi\, d_{CCH}\tan(\theta)\,f_y\right]} \qquad (10\text{-}19)$$

$MTF_{TDI}(f_y)$ is simply the MTF of an averaging filter for N_{TDI} samples displaced $d_{CCH}\tan(\theta)$ one relative to the next. Figure 10-12 illustrates the vertical MTF degradation due to image rotation for 32 and 96 stages. The spatial frequency is normalized to $d_{CCH}f_y$ where d_{CCH} is the *horizontal* detector pitch.

TDI offers the advantage that the signal-to-noise ratio increases by the square root of the number of TDI stages. This suggests that many TDI detectors should be used. The disadvantage is that the scan velocity must be tightly controlled as the number of TDI stages increases. System errors in scan velocity can dramatically decrease the MTF.

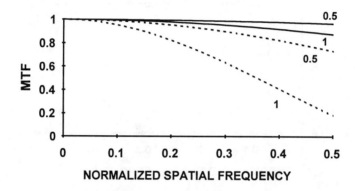

Figure 10-12. MTF degradation in the readout direction due to image rotation of 0.5 and 1 degrees with respect to the array axis. The spatial frequency axis is normalized to $d_{CCH}f_y$. The solid lines represent $N_{TDI} = 32$ and the dashed lines are for $N_{TDI} = 96$.

10.3.4. OPTICAL ANTI-ALIAS FILTER

In color filter arrays, like-colored sensitive detector center-to-center spacings may be much larger than the detector pitch. This creates a low fill-factor for each color and potentially increases aliasing (see Section 8.3.2., *Detector Array Nyquist Frequency*, page 209). The optical anti-alias filter blurs the image and effectively increases the optical area[12] of the detector. The filter, which consists of one or more birefringent crystals, is placed between the lens assembly and the detector array. Birefringent crystals break a beam into two components: the ordinary and the extraordinary. As illustrated in Figure 10-13, the detector collects light from a larger beam. Rays that would have missed the detector are refracted onto the detector. This makes the detector appear optically larger. While this changes the system MTF, the system sensitivity does not change. The light that is lost to adjoining areas is replaced by light refracted onto the detector.

The filter design depends upon the CFA design. If detectors sensitive to like colors are next to each other, the filter may not be required. If the detectors alternate, then, an appropriately designed filter will make the detectors appear twice as large. Similarly if the detectors are spaced three apart, then the filter will make the detector appear three times greater. This larger value, (the effective detector size) is used in Equation 10-12 (page 240).

SYSTEM MTF 249

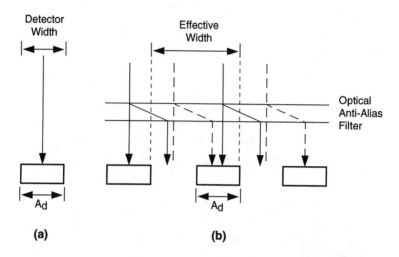

Figure 10-13. The birefringent filter increases the effective optical area of the detector. (a) No filter and (b) filter. Rays that would fall between detectors are refracted onto the detector.

Multiple crystals are used to change the effective detector size both vertically and horizontally. Since the spacing depends upon the CFA design, there is no unique anti-alias filter design. Clearly, if the detectors are unequally spaced, the Nyquist frequency for each color is different. Similarly, with different effective sizes, the MTFs are also different for each color. Complex algorithms are required to avoid color aliasing (see Example 5-1, page 141) and to equalize the MTFs. The anti-alias filter is essential for CFAs to reduce color Moiré patterns and can be used in any camera to reduce aliasing.

Figure 10-14 illustrates how the detector MTF is modified by the birefringent crystal when the detector pitch is twice as large as the element size. While the filter reduces aliasing, it also reduces the overall system MTF. For scientific applications, this reduction may be too severe.

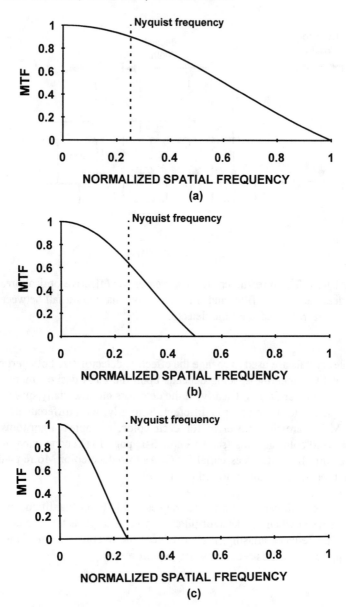

Figure 10-14. MTF$_{DETECTOR}$ modifications due to the anti-alias filter as a function df$_i$. The detector pitch is twice the detector element size. (a) No filter, (b) a filter that doubles the detector optical size, and (c) a filter that quadruples the effective detector size. The filter does not affect the array Nyquist frequency.

SYSTEM MTF 251

10.4. MOTION

Motion blurs imagery. When motion is fast compared to the CCD integration time, details become blurred in a reproduced image. In real-time imagery, edges may appear fuzzy to the observer even though any one frame may provide a sharp image. The eye integration time blends many frames. The effects of motion during the entire integration and human interpretation process must be considered.

Linear motion includes both target movement (relative to the imaging system) and motion of the imaging system (stationary target). High-frequency random motion is simply called jitter. Typically, an imaging system is subjected to linear and random motion simultaneously:

$$MTF_{MOTION} = MTF_{LINEAR} MTF_{RANDOM} \quad (10\text{-}20)$$

In a laboratory, the system and test targets may be mounted on a vibration-isolated stabilized table. Here, no motion is expected so that $MTF_{MOTION} = 1$.

10.4.1. LINEAR MOTION

The MTF degradation due to horizontal linear motion is:

$$MTF_{LINEAR}(f_x) = sinc(a_L f_x) \quad (10\text{-}21)$$

where a_L is the distance the target edge moved across the image plane. It is equal to $v_R \Delta t$. v_R is the relative image velocity between the sensor and the target. For scientific and machine vision applications, the exposure time is equal to the detector integration time. An observer's eye, on the other hand, blends many frames of data and the eye integration time should be used. Although the exact eye integration time is debatable, the values most often cited are between 0.1 and 0.2 s.

Figure 10-15 illustrates the MTF due to linear motion as a fraction of the detector size, d. Linear motion only affects the MTF in the direction of the motion. As a rule-of-the-thumb, when the linear motion causes an image shift less than about 20% of the detector size, it has minimal effect on system performance.

252 CCD ARRAYS, CAMERAS, and DISPLAYS

As the motion increases, a phase reversal occurs which is represented by a negative MTF. Here, periodic black bars appear white and periodic white bars appear black. Figure 10-16 illustrates a test pattern smeared by linear motion. At cutoff (where the system MTF approaches the first zero), the bar pattern cannot be resolved. Phase reversal occurs above cutoff. The ability to see imagery above cutoff is called false resolution or spurious resolution.

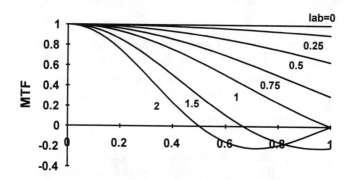

Figure 10-15. MTF_{LINEAR} as a function of normalized spatial frequency df_i. The curves represent values of a_L/d. In the laboratory a_L is usually zero. The negative MTF represents phase reversal.

10.4.2. RANDOM MOTION (JITTER)

With high-frequency motion, it is assumed that the image has moved often during the integration time so that the central limit theorem is valid. The central limit theorem says that many random movements can be described by a Gaussian distribution. The Gaussian MTF is

$$MTF_{RANDOM}(f_i) = e^{-2\pi^2 \sigma_R^2 f_i^2} \quad (10\text{-}22)$$

σ_R is the rms random displacement in mm. Figure 10-17 illustrates the random motion MTF. As a rule-of-thumb, when the rms value of the random motion is less than about 10% of d (detector size), system performance is not significantly affected. Jitter is considered equal in all directions and is included in both the horizontal and vertical system MTFs. High frequency motion blurs imagery (Figure 10-18).

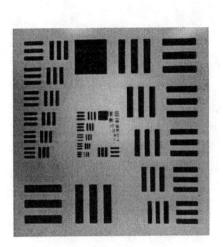
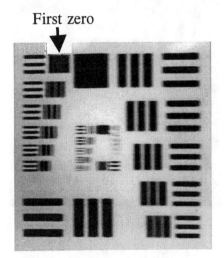

ORIGINAL IMAGE   LINEAR MOTION

Figure 10-16. Linear motion smears imagery. Since the motion is horizontal, the vertical resolution is maintained. When the bar target frequency is equal to the first zero of MTF_{LINEAR}, the bars all blend together. Higher spatial frequency bars illustrate phase reversal: There appear to be two black bars on a gray background. (From reference 13: by courtesy of G. L. Conrad).

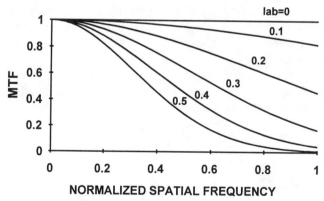

Figure 10-17. MTF degradation due to high-frequency random (Gaussian) movement as a function of $\sigma_R f_i$. In the laboratory, σ_R is usually zero.

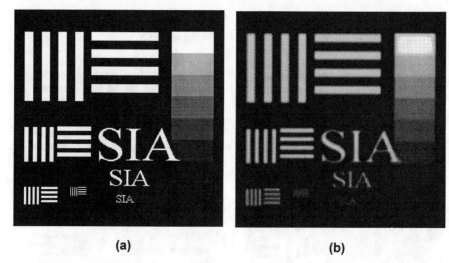

Figure 10-18. Random motion blurs imagery. (a) Original image, (b) two-dimensional random motion. Generated by SIA[14].

10.5. ELECTRONIC FILTERS

Appropriately designed electronic filters will not degrade the signal but will minimize the total noise. Boost filters may be used to enhance the signal at selected spatial frequencies. Both analog and digital filters can be used. Digital filters process data that resides in a memory. The units assigned to the filter match the units assigned to the data arrays. For convenience, the filter sampling frequency is the same as the array sampling frequency. With this mapping, each filter coefficient processes one pixel value. Digital filters can operate in either the vertical or horizontal directions.

Analog filters are causal and one-dimensional and modify a serial data stream. For analog filters immediately following the CCD array, the electrical frequency is related to the total number of detectors and the frame rate, F_R:

$$f_{elec} = \frac{N_{DETECTORS}}{F_R} \quad Hz \quad (10\text{-}23)$$

The sampling rate is related to the pixel clock by

$$f_{s-elec} = \frac{1}{t_{CLOCK}} \quad (10\text{-}24)$$

These filters operate on the data as it is read off the array. They apply to the direction of readout.

After the digital-to-analog converter, the serial stream data rate, and therefore the filter design, is linked to the video standard (see Section 5.2.1., *Video Timing*, page 127). These filters usually operate on a horizontal data stream. The active line time (Table 5-1, page 128) and array size provide the link between image space and video frequencies.

$$f_v = \frac{N_H d_{CCH}}{t_{LINE}} f_x \quad (10\text{-}25)$$

N_H is the number of horizontal detector elements with detector pitch d_{CCH} and t_{LINE} is the active video line time.

The electronic filter response is symbolically denoted by $H(f_{elec})$. When referred to image space it is labeled as the MTF. Ideally, $H(f_{elec})$ is a filter whose MTF is unity up to the array Nyquist frequency and then drops to zero. This filter maximizes the SNR by attenuating out-of-band amplifier noise and passing the signal without attenuation. The ideal filter is, of course, unrealizable but can be approximated by a high-order Butterworth filter:

$$H(f) = \frac{1}{\sqrt{1 + \left(\frac{f_{elec}}{f_{3db}}\right)^{2N}}} \quad (10\text{-}26)$$

where f_{3db} is the frequency at which the power is one-half. When f_{3db} is approximately equivalent to the array Nyquist frequency, the sampling theorem (with the appropriate reconstruction) is satisfied. However, square waves (bar targets) will appear as sinusoids (see Section 8.3.3., *Image Distortion*, page 212). Thus, f_{3db} must be much higher to preserve the square wave response.

10.5.1. DIGITAL FILTERS

There are many available image processing algorithms. Only a few of these can be described mathematically in closed-form and only these can be included in an end-to-end system performance model. The performance of the remaining algorithms can be inferred only by viewing the system output for a few representative inputs. Since the system is not spatially invariant, predicting system performance when viewing any other target is pure conjecture.

Digital filters provide any variety of pass bands to modify the frequency features of a digital data array. Two-dimensional filters are considered separable (i.e., the vertical and horizontal operations are independent). The horizontal filter affects only the system horizontal MTF and the vertical filter only affects the system vertical MTF.

The sampling process replicates the frequency spectrum at nf_S (see Section 8.1., *Sampling Theorem*, page 202). Similarly, digital filter response is symmetrical about the Nyquist frequency and repeats at multiples of the sampling frequency. The highest frequency of interest is the Nyquist frequency.

There are two general classes of digital filters[15]: infinite impulse response (IIR) and finite impulse response (FIR). Both have advantages and disadvantages. The FIR has a linear phase shift and the IIR does not. IIR filters tend to have excellent amplitude response whereas FIR filters tend to have more ripple. FIR filters are typically symmetrical in that the weightings are symmetrical about the center sample. They are also the easiest to implement in hardware or software. Figure 10-19 illustrates two FIR filters. The central data point is replaced by the weighted sum of neighboring data points. The filter is then moved one data point and the operation is repeated until the entire data set has been processed. There exists edge effects with any digital filter. The filter illustrated in Figure 10-19a requires seven inputs before a valid output can be achieved. At the very beginning of the data set, there are insufficient data points to have a valid output at data point 1, 2, or 3. The user must be aware of edge effects at both the beginning and the end of the data record. This means that image periphery cannot be filtered. The following MTF equations are only valid in the central portion of the field-of-view. For FIR filters where the multiplicative factors (weightings) are symmetrical about the center, the filter is mathematically represented by a cosine series (sometimes called a cosine filter).

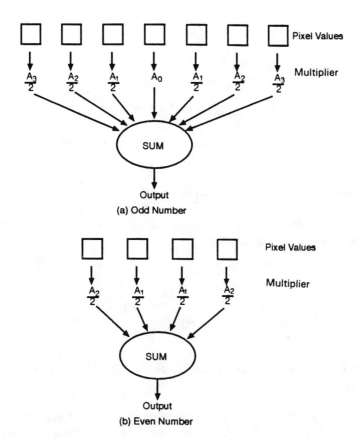

Figure 10-19. Symmetrical digital filters. (a) 7-tap (odd number) filter and (b) 4-tap (even number) filter.

For an odd number of samples (also called taps) summed,

$$MTF_{DFILTER}(f_i) = \sum_{k=0}^{\frac{N-1}{2}} A_k \cos\left(\frac{2\pi k f_i}{f_S}\right) \quad (10\text{-}27)$$

For an even number of samples,

$$MTF_{DFILTER}(f_i) = \sum_{k=1}^{\frac{N}{2}} A_k \cos\left(\frac{\pi(2k-1)f_i}{f_S}\right) \quad (10\text{-}28)$$

The sum of the coefficients should equal unity so the MTF is one at $f_x = 0$

$$\sum A_k = 1 \quad (10\text{-}29)$$

f_S is the array sampling frequency ($f_S = 1/d_{CC}$).

With an averaging filter[10], all the multipliers shown in Figure 10-19 are equal. It can be represented by,

$$MTF_{DFILTER}(f_i) = \frac{\sin\left(N_{AVE}\pi\frac{f_i}{f_S}\right)}{N_{ave}\sin\left(\pi\frac{f_i}{f_S}\right)} = \frac{\text{sinc}\left(N_{AVE}\frac{f_i}{f_S}\right)}{\text{sinc}\left(\frac{f_i}{f_S}\right)} \quad (10\text{-}30)$$

N_{AVE} is the number of pixels averaged together. The averaging filter has its first zero at $f_x = f_S/N_{AVE}$.

Averaging samples together is called binning or super pixeling. It provides the same MTF as if the detector elements were proportionally larger. That is, averaging two samples together provides the same MTF as a single detector element that is twice as large. Cameras that operate in either pseudo-interlace and progressive scan modes will have a different vertical MTF for each mode. In the pseudo interlace mode, two pixels are averaged together.

If $d = d_{CC}$, then $MTF_{DETECTOR}MTF_{DFILTER}$ is

$$sinc(df_i) \frac{sinc\left(N_{AVE}\frac{f_i}{f_S}\right)}{sinc\left(\frac{f_i}{f_S}\right)} = sinc(N_{AVE}df_i) \quad (10\text{-}31)$$

where the effective detector width is $N_{AVE}d$.

10.5.2. SAMPLE-AND-HOLD

After the digital-to-analog converter (DAC), the analog signal changes only at discrete times. The sample-and-hold circuitry within the DAC extends the data into a continuous analog signal. The MTF of a zero-order sample-and-hold is:

$$MTF_{S\&H} = sinc\left(\frac{f_x}{f_S}\right) \quad (10\text{-}32)$$

where $f_S = 1/d_{CC}$. Figure 10-20 illustrates the sample-and-hold MTF as a function of normalized spatial frequency.

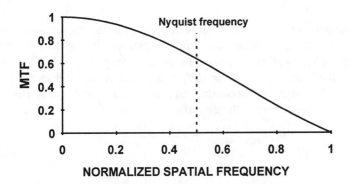

Figure 10-20. Zero-order sample-and-hold MTF normalized to $d_{CC}f_i$. The sample-and-hold acts as a low-pass filter.

10.5.3. POST-RECONSTRUCTION FILTER

After the digital-to-analog conversion, the image appears still blocky due to the discrete nature of the conversion. The ideal post-reconstruction filter removes all the higher order frequencies (Figure 10-21) such that only the original smooth signal remains. The output is delayed but this not noticed on the imagery.

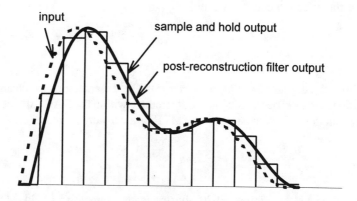

Figure 10-21. The post-reconstruction filter removes the blocky (stair step) effect created by the sample-and-hold circuitry within the DAC.

An ideal post-reconstruction filter will only pass the signal information contained in the originally sampled bandwidth. For an ideal system, the post-reconstruction filter MTF is unity up to the Nyquist frequency. Real filters will have some roll-off and f_{3db} should be sufficiently large so that it does not affect the in-band MTF. Filters can be constructed in many functional forms. The most common filter is the Nth-order Butterworth:

$$MTF_{POST-RECON} = \frac{1}{\sqrt{1 + \left(\frac{f_v}{f_{3db}}\right)^{2N}}} \qquad (10\text{-}33)$$

Example 10-2
POST-RECONSTRUCTION FILTER

What is the required spatial frequency cutoff for an ideal post-reconstruction filter if the desired output is EIA 170 and there are 640 horizontal detectors in the array?

For EIA 170, t_{LINE} is 52.09 µs. The Nyquist frequency in the video domain is

$$f_{VN} = \frac{1}{2} \frac{640}{52.09 \times 10^{-6}} = 6.14 \; Mhz \qquad (10\text{-}34)$$

The bandwidth necessary to transmit the detector data must exceed 6.14 MHz to avoid roll-off effects. This is a greater bandwidth than allowed by the EIA 170 standard. However, the camera output only has to conform to EIA 170 timing so that the information can be displayed on a conventional monitor.

10.5.4. BOOST

MTF degradation caused by the various subsystems can be partially compensated with electronic boost filters. Boost filters also amplify noise so that the signal-to-noise ratio may degrade. For systems that are contrast limited, boost may improve image quality. However, for noisy images, the advantages of boost are less obvious. As a result, these filters are used only in high-contrast situations (typical of consumer applications) and are not used in scientific applications where low signal-to-noise situations are often encountered. Historically, each subsystem was called an "aperture." Therefore boost provided "aperture correction." This should not be confused with the optics MTF where the optical diameter is the entrance aperture.

The boost amplifier can either be an analog or digital circuit whose peaking compensates for any specified MTF roll-off. The MTF of a boost filter, by definition, exceeds one over a limited spatial frequency range. When used with the all the other subsystem MTFs, the resultant MTF_{SYS} is typically less than one for all spatial frequencies. Excessive boost can cause ringing at sharp edges. Horizontal filters create ringing in the horizontal direction only.

A variety of boost circuits are available. One example is the tuned circuit whose response is

$$H_{BOOST}(f_v) = \frac{1}{\sqrt{\left(1 - \left(\frac{f_v}{f_{BOOST}}\right)^2\right)^2 + \left(\frac{f_v}{Q f_{BOOST}}\right)^2}} \qquad (10\text{-}35)$$

Q is the quality factor and equal to boost amplitude when $f_v = f_{BOOST}$.

Although the boost is illustrated in the video domain, boost can be placed in any part of the circuit where a serial stream of data exits. By appropriate selection of coefficients (see Equations 10-27 and 10-28, page 260), digital filters can also provide boost. MTF_{BOOST} can also approximate the inverse of MTF_{SYS} (without boost). The combination of MTF_{BOOST} and MTF_{SYS} provides unity MTF over spatial frequencies of interest. This suggests that the reproduced image will precisely match the target in every spatial detail.

10.6. DISPLAY

The display MTF is a composite MTF that includes both the internal amplifier response and the CRT response. Implicit in the MTF is the conversion from input voltage to output display brightness. Although not explicitly stated, the equation implies radial symmetry: The MTF is the same in both the vertical and horizontal directions. While this may not be precisely true, the assumption is adequate for most calculations. It is reasonable to assume that the resultant spot is Gaussian (see Section 7.3., *Spot Size*, page 178). Then

$$MTF_{DISPLAY}(f_{raster}) \approx e^{-2\pi^2 \sigma_{SPOT}^2 f_{raster}^2} = e^{-2\pi^2 \left(\frac{S}{2.35} f_{raster}\right)^2} \qquad (10\text{-}36)$$

f_{RASTER} is the raster frequency in units of cy/mm on the monitor screen, σ_{SPOT} is the standard deviation of the Gaussian beam profile in mm, and S is the spot diameter (FWHM) in mm.

SYSTEM MTF

Assuming radial symmetry and that the display has the same aspect ratio as the camera, the standard deviation in image space is

$$\sigma_{IS} = \frac{S}{2.35} \frac{N_H d_{CCH}}{W_{MONITOR}} = \frac{S}{2.35} \frac{N_V d_{CCV}}{H_{MONITOR}} \quad mm \quad (10\text{-}37)$$

$W_{MONITOR}$ ($H_{MONITOR}$) is the physical width (height) of the CRT measured in the same units as the spot size. N_H and N_V are the number of detectors in the horizontal and vertical directions. Similarly d_{CCH} and d_{CCV} are the detector pitches. The MTF is

$$MTF_{DISPLAY}(f_i) = e^{-2\pi^2 \left(\frac{S}{2.35} \frac{N_H d_{CCH}}{W_{MONITOR}} f_i\right)^2} = e^{-2\pi^2 \left(\frac{S}{2.35} \frac{N_V d_{CCV}}{H_{MONITOR}} f_i\right)^2} \quad (10\text{-}38)$$

If the spot size is not specified, then σ_{SPOT} can be estimated from the TV limiting resolution (see Section 7.5.3, *TV Limiting Resolution*, page 187). Each line in the test pattern is assumed to be separated by $1.18\sigma_{SPOT}$. Since there are two lines per cycle, each cycle is $2.36\sigma_{SPOT}$ wide. If N_{TV} lines per picture height are displayed on a monitor whose aspect ratio is α_m, $W_{MONITOR} \approx 2.35\sigma_{SPOT}\alpha_m N_{TV}$ and $H_{MONITOR} \approx 2.35\sigma_{SPOT}N_{TV}$. Since $S = 2.35\sigma_{SPOT}$, $H_{MONITOR} \approx S N_{TV}$, then for the flat field condition:

$$MTF_{DISPLAY}(f_i) \approx e^{-2\pi^2 \left(\frac{N_V d_V}{2.35 N_{TV}} f_i\right)^2} \quad (10\text{-}39)$$

Figure 10-22 illustrates $MTF_{DETECTOR}$ and $MTF_{DISPLAY}$. The display spot size is independent of the detector size and the display resolution is independent of the number of detector elements. Matching the display resolution (number of pixels) to the number of detector elements does not uniquely specify the display MTF. For example, suppose an array contained 1000 elements horizontally. A monitor that can resolve 1000 pixels may not provide good imagery if the resolution/addressability ratio (RAR) is large. The RAR was described in Section 7.6., *Addressability* (page 189).

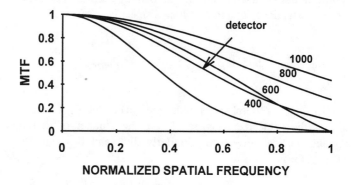

Figure 10-22. MTF$_{DISPLAY}$ as a function of $d_{CCV}f_i$ for four different TV limiting resolutions when $N_V = 480$. For most systems, the display MTF limits the system MTF. The Nyquist frequency (not shown) is at $d_{CCV}f_i = 0.5$.

☞

Example 10-3
DISPLAY MTF

What should the display resolution be to maximize the system MTF?

In Example 7-2 (page 194), the RAR was calculated for a variety of TV limiting resolutions. The resultant MTF (Figure 7-12, page 192) applies to the visibility of point sources. Equation 10-38 or Equation 10-39 is used to calculate image MTF when the input is a sinusoid. The Nyquist frequency is used most often when specifying system response. For EIA 170, $N_V = 485$. Inserting $d_V f_i = 0.5$ into Equation 11-19 provides

$$MTF_{DISPLAY}(f_N) = e^{-2\pi^2 \left(\frac{0.5}{2.35} \frac{485}{N_{TV}}\right)^2} \qquad (10\text{-}40)$$

For a 100% fill-factor array, the detector MTF is 0.63 at f_N. Figure 10-23 illustrates the detector and display MTF at Nyquist frequency. The display resolution must increase to 1400 TVL/PH for display MTF to be 0.90 at the Nyquist frequency. This suggests that most imaging systems are display resolution limited. This result is based upon the assumption that the beam profile is Gaussian. A different shaped beam (i.e., flat-topped), will provide a different relationship between TV limiting resolution and MTF.

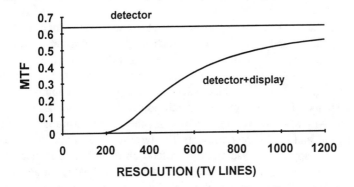

Figure 10-23. $MTF_{DETECTOR} MTF_{DISPLAY}$ at Nyquist frequency as a function of TV liming resolution. The detectors are contiguous so that the center-to-center spacing is equal to the detector element size. $N_V = 485$. All other MTFs are assumed to be unity at f_N.

10.7. EYE RESPONSE

The observer processes imagery to create a perceived MTF:

$$MTF_{PERCEIVED} = MTF_{SYS} MTF_{EYE} \quad (10\text{-}41)$$

The eye MTF should only be used to understand the relationship between the imaging system MTF and the eye response. It is not used to describe sensor performance. When $MTF_{PERCEIVED}$ is coupled to system noise, system responsivity, and the eye integration capabilities, it is possible to predict the minimum resolvable contrast (described in Chapter 12, *Minimum Resolvable Contrast*). The eye's contrast sensitivity is also included in all image quality metrics (discussed in Chapter 11, *Image Quality*).

The sine wave response is used as an approximation to the eye-brain MTF. It is sometimes called the human visual system MTF or HVS-MTF. The sine wave response depends on diffraction by the pupil, aberrations of the lens, finite size of the photoreceptors, ocular tremor, and neural interconnections within the retina and brain. Diffraction and aberrations vary with ambient luminance, monitor brightness and chromatic composition of the light. Since the retina is composed of rods and cones of varying densities, the location and size of the object with respect to the fovea significantly affect the sine wave response.

The sine wave response ignores spatial noise, background luminance, angular orientation and exposure time. Each of these parameters significantly affects the interpretation of image visibility. The MTF in the purest sense is noise independent but the human eye's response is very sensitive to spatial and temporal noise. Therefore, the sine wave response is only an approximation to the true response. Furthermore, the overall population exhibits large variations in response. Any MTF approximation used for the eye therefore is only a crude approximation and probably represents the largest uncertainty in the overall MTF analysis approach. Factors that affect observer performance include monitor viewing ratio (usually selected to reduce raster effects), target size, and observation time. Visual psychophysical data can be found in Farrell and Booth's[16] and Biberman's[17] books.

The usual measurement is to determine the minimum modulation required to just perceive a sinusoidal target. The modulation is

$$M_t = \frac{L_{MAX} - L_{MIN}}{L_{MAX} + L_{MIN}} \qquad (10\text{-}42)$$

where L_{MAX} and L_{MIN} are the maximum and minimum luminances respectively (see Figure 8-1, page 199). The eye's detection capability depends upon the visual angle subtended by the target size at the observer. As shown in Figure 10-24, in the absence of noise, the eye's modulation threshold (minimum perceivable modulation) is characteristically J-shaped. It is labeled as M_t in the visual psychophysical literature. The eye is most sensitive to spatial frequencies[18] that range between 3 and 5 cy/deg at typical ambient lighting levels.

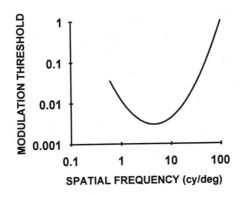

Figure 10-24. A representative modulation threshold function.

The increase in threshold at low frequencies is due to the eye's inhibitory signal processing component. The contrast sensitivity is the inverse of the modulation threshold curve. When normalized to one, the contrast sensitivity or sine wave response is called the eye MTF. By convention, the MTF is defined to be one at zero spatial frequency and to be a monotonically decreasing function. The way to include the inhibitory response as an MTF component is not clear.

Nevertheless, various researchers modeled the normalized contrast sensitivity (Figure 10-25) and labeled it as the eye MTF. Nill[19] recommended:

$$MTF_{EYE}(f_{eye}) = 2.71\left[0.19 + 0.81\left(\frac{f_{eye}}{f_{PEAK}}\right)\right]e^{-\left(\frac{f_{eye}}{f_{PEAK}}\right)} \quad (10\text{-}43)$$

Schulze[20] recommended

$$MTF_{EYE}(f_{eye}) = 2.71\left(e^{-0.1138 f_{eye}} - e^{-0.325 f_{eye}}\right) \quad (10\text{-}44)$$

Campbell-Robson's high luminance data[21] can be approximated by

$$MTF_{EYE}(f_{eye}) = 10^{-2.8 \log_{10}\left(\frac{f_{eye}}{f_{PEAK}}\right)} \quad (10\text{-}45)$$

deJong and Bakker[22] used

$$MTF_{EYE}(f_{eye}) = \sin^2\left[\frac{\pi}{2}\left(\frac{f_{eye}}{f_{PEAK}}\right)^{1/3}\right] \quad (10\text{-}46)$$

Neglected the low-frequency inhibitory response, Kornfeld and Lawson[23] suggested that

$$MTF_{EYE}(f_{eye}) = e^{-\Gamma\frac{f_{eye}}{17.45}} \quad (10\text{-}47)$$

Γ is a light-level dependent eye response factor that was presented in tabular form[24] by Ratches et. al. A third order polynomial fit provides

$$\Gamma = 1.444 - 0.344\log(B) + 0.0395\log^2(B) + 0.00197\log^3(B) \quad (10\text{-}48)$$

B is the monitor brightness in foot-lamberts.

Considering the large variation in observer responses, these curves have a similar shape when $f_{eye} > f_{PEAK}$. The original NVL static performance model[24] and FLIR92[25] use the Kornfeld-Lawson eye model. These models were validated by the US Army when detecting tank-sized targets at modest ranges. This translates into modest spatial frequencies. The Kornfeld-Lawson model appears to follow all other models (to within a multiplicative factor) for mid-range spatial frequencies (Figure 10-25). This multiplicative factor becomes a normalization issue for the MRC (discussed in Chapter 12, *Minimum Resolvable Contrast*).

Insufficient data exists to say with certainty which eye model is best. Because of this uncertainty, *all analyses, no matter which eye model is used, must only be used for comparative performance purposes.*

Two scenarios are possible: (1) the observer is allowed to move his head and (2) the head is fixed in space. Since the eye's detection capability depends upon the angular subtense of the target, head movement may provide different results than if the head is fixed. In laboratory testing, the distance to the monitor is not usually specified nor limited in any way. To maximize detection capability (stay on the minimum of the contrast threshold curve), an observer subconsciously moves toward the monitor to perceive small targets and further away to see larger targets.

By allowing the observer to adjust the viewing distance, several interrelated detection criteria are optimized that include striving for apparent edge sharpness and maximizing perceived signal-to-noise ratio. This apparently results in an equal detection capability for all spatial frequencies such that eye's contrast sensitivity approaches a constant. This results in a nearly constant MTF that is called the "non-limiting eye MTF" or $MTF_{EYE} = 1$. Equation 10-48 is used in those situations where the head is fixed such as a pilot strapped into a cockpit seat.

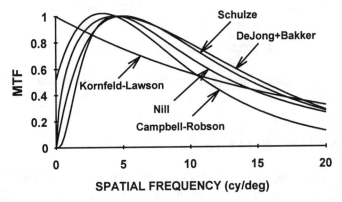

Figure 10-25. Various mathematical equations that approximate the eye's MTF.

10.8. INTENSIFIED CCD

The intensified CCD (ICCD) is more difficult to analyze. The MTF of an image intensifier tube (IIT) varies with tube type[26], manufacturing precision[27], and materials used. Figure 10-26 illustrates a possible range of MTFs. The MTF should be obtained from the manufacturer before proceeding with system analysis. It can be approximated by a simple Gaussian expression:

$$MTF_{INTENSIFIER}(f_x) = e^{-2\pi^2 \sigma_{IIT}^2 f_x^2} \qquad (10\text{-}49)$$

MTF theory will provide an optimistic result of fiber optic coupled ICCDs. It does not include the CCD mismatch or the chicken wire effects. The MTF associated with lens coupling is probably more accurate. The MTF of an ICCD is $MTF_{INTENSIFIER}$ multiplied by the appropriate MTFs presented in this chapter.

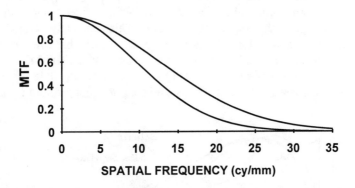

Figure 10-26. Range of MTFs for GenII image intensifier (From reference 27).

10.9. SAMPLING EFFECTS

For a sampled data system, a wide range of MTF values is possible[28-31] for any given spatial frequency (see Section 8.3.3, *Image Distortion*, page 212). To account for this variation, a sample-scene MTF may be included. The detector spatial frequency response is represented by $MTF_{DETECTOR}MTF_{PHASE}$. This MTF provides an average performance response that may be used to calculate the system response general imagery.

In general, the "MTF" is[32]

$$MTF_{PHASE}(f_x) \approx \cos\left(\frac{f_x}{f_N}\theta\right) \qquad (10\text{-}50)$$

where θ is the phase angle between the target and the sampling lattice. For example, when $f_x = f_N$, the MTF is a maximum when $\theta = 0$ (in-phase) and a zero when $\theta = \pi/2$ (out-of-phase).

To approximate a median value for phasing, θ is set to $\pi/4$. Here, approximately one-half of the time the MTF will be higher and one-half of the time the MTF will be lower. The median sampling MTF is

$$MTF_{MEDIAN}(f_x) \approx \cos\left(\frac{\pi}{2} \frac{f_x}{f_s}\right) \qquad (10\text{-}51)$$

At Nyquist frequency, MTF_{PHASE} is 0.707. This is the Kell factor so often reported when specifying the resolution of monitors. An MTF averaged over all phases is represented by

$$MTF_{AVERAGE}(f_x) \approx \text{sinc}\left(\frac{f_x}{f_s}\right) \qquad (10\text{-}52)$$

Figure 10-27 illustrates the difference between the two equations. Since these are approximations, they may be considered roughly equal over the range of interest (zero to the Nyquist frequency). MTF_{PHASE} can be applied to both the vertical and horizontal MTFs.

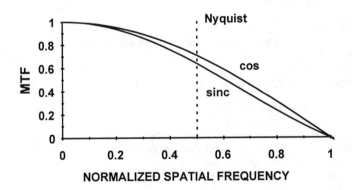

Figure 10-27. Average and median scene-sample phase MTFs normalized to f_x/f_s. The MTF is defined only up to the Nyquist frequency.

Sampling effects are phase dependent and, as a result, has no unique MTF. For system modeling purposes, average MTFs were assigned for some more common image processing algorithms. Because average MTFs are used, only an average system performance can be predicted. Performance can vary lattice. For laboratory measurements, it is common practice to "peak-up" the target (in-phase relationship) and then $MTF_{PHASE} = 1$. This is done to get repeatable results.

Frequencies above f_N are aliased down to lower frequencies. Signal aliasing is ignored in most analyses and has been ignored here. A complete end-to-end analysis should include both aliased noise and aliased signal[33]. These aliased components limit[34] the extent to which a sampled image can be sharpened. Image reconstruction may include data removal (decimation) or data expansion (interpolation). The effective sampling frequency and aliasing can appear to change using different image processing algorithms[35-38].

10.10. REFERENCES

1. See, for example, R. J. Farrell and J. M. Booth, *Design Handbook for Imagery Interpretation Equipment*, Reprinted with corrections, Report D180-19063-1, Boeing Aerospace Company, Seattle Wash, February 1984.
2. G. C. Holst, *Testing and Evaluation of Infrared Imaging Systems*, pp. 106-108, JCD Publishing, Winter Park, Fl (1993).
3. G. C. Holst, *Testing and Evaluation of Infrared Imaging Systems*, pp. 195-200, JCD Publishing, Winter Park, Fl (1993).
4. G. C. Holst, *Testing and Evaluation of Infrared Imaging Systems*, pp. 104-105, JCD Publishing, Winter Park, Fl (1993).
5. H.H. Hosack, "Aperture Response and Optical Performance of Patterned-Electrode Virtual-Phase Imagers," *IEEE Transactions on Electron Devices*, Vol. ED-28(1) pp. 53-63 (1981).
6. D. H. Sieb, "Carrier Diffusion Degradation of Modulation Transfer Function in Charge Coupled Imagers," *IEEE Transactions on Electron Devices*, Vol. ED-21(5), pp. 210-217 (1974).
7. R. H. Dyck, "VLSI Imagers," in *VLSI Electronics: Microstructure Science, Volume 5*, N. G. Einspruch, ed., pp. 594, Academic Press, New York (1985).
8. L. W. Schumann and T. S. Lomheim, "Modulation Transfer Function and Quantum Efficiency Correlation at Long Wavelengths (Greater Than 800 nm) in Linear Charge Coupled Imagers," *Applied Optics*, Vol. 28(9), pp. 1701-1709 (1989).
9. E. L. Dereniak and D. G. Crowe, *Optical Radiation Detectors*, pp.199-203, John Wiley & Sons, New York (1984).
10. H. V. Kennedy, "Miscellaneous Modulation Transfer Function (MTF) Effects Relating to Sampling Summing," in *Infrared Imaging Systems: Design, Analysis, Modeling, and Testing*, G. C. Holst, ed., SPIE Proceedings Vol. 1488, pp. 165-176 (1991).
11. H.-S Wong. Y. L. Yao, and E. S. Schlig, "TDI Charge-coupled Devices: Design and Applications," *IBM Journal of Research and Development*, Vol. 36(1), pp 83-105 (1992).
12. J. E. Greivenkamp, "Color Dependent Optical Prefilter for the Suppression of Aliasing Artifacts," *Applied Optics*, Vol. 29(5), pp 676-684 (1990).
13. G. Conrad, "Reconnaissance System Performance Predictions Through Image Processing," in *Airborne Reconnaissance XIV*, P A. Henkel, F. R. LaGesse, and W. W. Schurter, eds., SPIE Proceedings Vol. 1342, pp. 138-145 (1990).

SYSTEM MTF 273

14. SIA (System Image Analysis) software is available from JCD Publishing, 2932 Cove Trail, Winter Park, FL 32789.
15. There exists a variety of texts on digital filter design. See, for example, *Digital Signal Processing*, A. V. Oppenheim and R. W. Schafer, Prentice-Hall, New Jersey (1975).
16. R. J. Farrell and J. M. Booth, *Design Handbook for Imagery Interpretation Equipment*, Reprinted with corrections, Report D180-19063-1, Boeing Aerospace Company, Seattle Wash, February 1984.
17. L. M. Biberman, ed., *Perception of Displayed Information*, Plenum Press, New York (1973).
18. B. O. Hultgren, "Subjective Quality Factor Revisited," in *Human Vision and Electronic Imaging: Models, Methods and Applications*, B. E. Rogowitz and J. P. Allebach, eds., SPIE Proceedings Vol. 1249, pp. 12-22 (1990).
19. N. Nill, "A Visual Model Weighted Cosine Transform for Image Compression and Quality Measurements," *IEEE Transactions on Communications*, Vol. 33(6), pp. 551-557 (1985).
20. T. J. Schulze, "A Procedure for Calculating the Resolution of Electro-Optical Systems," in *Airborne Reconnaissance XIV*, P A. Henkel, F. R. LaGesse, and W. W. Schurter, eds., SPIE Proceedings Vol. 1342, pp. 317-327, (1990).
21. F. W. Campbell and J. G. Robson, "Application of Fourier Analysis to the Visibility of Gratings," *Journal of Physiology*, Vol. 197, pp. 551-566 (1968).
22. A. N. deJong and S. J. M. Bakker, "Fast and Objective MRTD Measurements," in *Infrared Systems - Design and Testing*, P. R. Hall and J. S. Seeley, eds., SPIE Proceedings Vol. 916, pp. 127-143 (1988).
23. G. H. Kornfeld and W. R. Lawson, "Visual Perception Model," *Journal of the Optical Society of America*, Vol. 61(6), pp. 811-820 (1971).
24. J. Ratches, W. R. Lawson, L. P. Obert, R. J. Bergemann, T. W. Cassidy, and J. M. Swenson, "Night Vision Laboratory Static Performance Model for Thermal Viewing Systems," US Army Electronics Command Report ECOM Report 7043, pg. 11, Ft. Monmouth, NJ (1975).
25. *FLIR92 Thermal Imaging Systems Performance Model*, Analyst's Reference Guide, Document RG5008993, Ft. Belvoir, VA, January 1993.
26. G. M. Williams Jr., "A High Performance LLLTV CCD Camera for Nighttime Pilotage," in *Electron Tubes and Image Intensifiers*, C. B. Johnson and B. N. Laprade, eds., SPIE Proceedings Vol. 1655, pp. 14-32 (1992).
27. I. P. Csorba, *Image Tubes*, pp.79-103, Howard W. Sams, Indianapolis, IN (1985).
28. S. K. Park and R. A. Schowengerdt, "Image Sampling, Reconstruction and the Effect of Sample-scene Phasing," *Applied Optics*, Vol. 21(17), pp. 3142-3151 (1982).
29. J. C. Feltz and M. A. Karim, "Modulation Transfer Function of Charge-coupled Devices," *Applied Optics*, Vol. 29(5), pp. 717-722 (1990).
30. J. C. Feltz, "Development of the Modulation Transfer Function and Contrast Transfer Function for Discrete Systems, Particularly Charge-coupled Devices," *Optical Engineering*, Vol. 29(8), pp. 893-904 (1990).
31. L. de Luca and G. Cardone, "Modulation Transfer Function Cascade Model for a Sampled IR Imaging System," *Applied Optics*, Vol. 30(13), pp. 1659-1664 (1991).
32. F. A. Rosell, "Effects of Image Sampling," in *The Fundamentals of Thermal Imaging Systems*, F. Rosell and G. Harvey, eds., pg. 217, NRL Report 8311, Naval Research Laboratory, Wash D.C. (1979).
33. S. K. Park, "Image Gathering, Interpolation and Restoration: A Fidelity Analysis," in *Visual Information Processing*, F. O. Huck and R. D. Juday, eds., SPIE Proceedings Vol. 1705, pp. 134-144 (1992).
34. S. K. Park and R. Hazra, "Image Restoration Versus Aliased Noise Enhancement," in *Visual Information Processing III*, F. O. Huck and R. D. Juday, eds., SPIE Proceedings Vol. 2239, pp. 52-62 (1994).

35. I. Ajewole, "A Comparative Analysis of Digital Filters for Image Decimation and Interpolation," in *Applications of Digital Image Processing VIII*, A. G. Tescher, ed., SPIE Proceedings Vol. 575, pp. 2-12 (1985).
36. W. F. Schrieber and D. E. Troxel, "Transformation Between Continuous and Discrete Representations of Images: A Perceptual Approach," *IEEE Transactions on Pattern Analysis and Machine Intelligence*, Vol. PAMA-7(2), pp. 178-186 (1985).
37. R. W. Schafer and L. R. Rabiner, "A Digital Signal Processing Approach to Interpolation," *Proceedings of the IEEE*, Vol. 61(8), pp. 692-702 (1973).
38. A. H. Lettington and Q. H. Hong, "Interpolator for Infrared Images," *Optical Engineering*, Vol. 33(3), pp. 725-729 (1994).

11
IMAGE QUALITY

Our perception of good image quality is based upon the real-world experiences of seeing all colors, all intensities and textures. An imaging system has a limited field-of-view, limited temporal and spatial resolutions, and presents a two-dimensional view of a three-dimensional world. In the real-world our eyes scan the entire scene. Not all of the available information is captured by an imaging system. Furthermore, an imaging system introduces noise and the loss of image quality due to noise can only be estimated. Cameras sensitive to wavelengths less than 0.4 μm and greater than 0.7 μm provide imagery that we cannot directly perceive. The quality of the ultraviolet or infrared imagery can only be estimated because we do not know how it really appears.

In some respects, the starting point for camera system design should be the monitor. No image will look sharp unless it is limited by the eye MTF. The best image quality only occurs when the observer is at an optimum viewing distance. If too far away, detail will be beyond the eye's resolution limit. At a large distance even poor imagery will appear good. As the observer approaches the display, a point is reached where the image does not become clearer because there is no further detail to see. Display size and resolution requirements are based on an assumed viewing distance.

Image quality is a subjective impression ranging from poor to excellent. It is a somewhat learned ability. It is a perceptual one, accomplished by the brain, affected by and incorporating inputs from other sensory systems, emotions, learning and memory. The relationships are many and are still not well understood. Perceptual quality of the same scene varies between individuals and temporally for the same individual. There exist large variations in an observer's judgment as to the correct rank ordering of image quality from poor to best and therefore image quality cannot be placed on an absolute scale. Visual psychophysical investigations have not measured all the properties relevant to imaging systems.

Many formulas exist for predicting image quality. Each is appropriate under a particular set of viewing conditions. These expressions are typically obtained from empirical data in which multiple observers view many images with a known amount of degradation. The observers rank the imagery from worst to best and then an equation is derived which relates the ranking scale to the amount of degradation.

If the only metric for image quality was resolution, then we would attempt to maximize resolution in our system design. Many tests have provided insight into image quality metrics. In general, images with higher MTFs and less noise are judged as having *better* image quality. There is no single *ideal* MTF shape that provides best image quality. For example, Kusaka[1] showed that the MTF that produced the most aesthetically pleasing images depended on the scene content.

The metrics suggested by Granger and Cupery[2], Shade[3], and Barten[4,5,6] offer additional insight on how to optimize an imaging system. Granger and Cupery developed the Subjective Quality Factor (SQF): an empirically derived relationship using individuals' responses when viewing many photographs. Shade used photographs and included high-quality TV images. Barten's approach is more comprehensive in that it includes a variety of display parameters. It now includes[7] contrast, luminance, viewing ratio, number of scan lines, and noise.

While Barten has incorporated the sampling effects of flat panel displays[7], no model includes the sampling (and associated aliasing) that takes place at the detector. Until this aliasing is quantified, no metric will predict image quality for sampled data systems. Only an end-to-end system approach can infer overall image quality.

There are potentially two different system design requirements: (1) Good image quality and (2) performing a specific task. Sometimes these are equivalent, other times they are not. All image quality metrics incorporate some form of the system MTF. The underlying assumption is that the image spectrum is limited by the system MTF. Equivalently, it is assumed that the scene contains all spatial frequencies and that the displayed image is limited by system MTF - a reasonable assumption for general imagery. While good image quality is always desired, a military system is designed to detect and recognize specific targets (discussed in Chapter 12, *Minimum Resolvable Contrast*). Optimized military systems will have high MTF at the (assumed) target frequencies and other spatial frequencies are considered less important. Computer monitors are usually designed so that alphanumeric characters are legible.

The eye-brain system appears to operate as a tuned spatial-temporal filter where the tuning varies according to the task on hand. Since the eye approximates an optimum filter, no system performance improvement is expected by *precisely* matching the image spectrum to the eye-brain preferred spectrum. Rather, if the displayed spectrum is within the limits of the eye spectrum, the eye will automatically tune to the image. Clearly the overall system magnification should be set such that the frequency of the maximum interest coincides with the peak frequency of the eye's contrast sensitivity. Although this implies a specific frequency, the range of optimization is broad.

A large number of metrics are related to image quality. Most are based on monochrome imagery such as resolution, MTF, and minimum resolvable contrast. Color reproduction and tonal transfer issues, which are very important to color cameras, are not covered here.

The symbols used in this book are summarized in the *Symbol List* (page xiv) and it appears after the *Table of Contents*.

11.1. RESOLUTION METRICS

An overwhelming majority of imaging quality discussions center on resolution or sensitivity. Resolution has been in use so long that it is thought to be fundamental and to uniquely determine system performance. It implies something about the smallest target detail that can be resolved. It may be specified by a variety of sometimes unrelated metrics such as the Airy disk angular size, the detector angular subtense (DAS), detector pitch, or the Nyquist frequency.

Resolution is one measure of image quality. Subconsciously, we think that systems with better resolution will provide better image quality. This is generally true for systems of similar designs with MTFs of the same functional form.

There are four different types of resolution: (1) temporal resolution, which is the ability to separate events in time, (2) gray scale resolution, which is determined by the analog-to-digital converter design, noise floor, or the monitor capability, (3) spectral resolution, and (4) spatial resolution. An imaging system operating at 30 Hz frame rate has a temporal resolution of 1/30 sec. Gray scale resolution is a measure of the dynamic range. The spectral resolution is simply the spectral band pass (e.g., UV, visible, or near infrared) of the system. This section covers spatial resolution.

278 CCD ARRAYS, CAMERAS, and DISPLAYS

Resolution provides valuable information regarding the finest spatial detail that can be discerned. Each discipline extracts its own type of information from data and each discipline has its own requirements for resolution (Table 11-1). A large variety of resolution measures exist and the various definitions may not be interchangeable. As imaging systems are incorporated into new disciplines, resolution must be specified in terms used by those industries. For example, photo interpreters use the ground resolved distance (GRD) when evaluating reconnaissance imagery captured on film. But as CCD-based imaging systems replace the wet film process, the GRD, by default, becomes an imaging system resolution measure.

Table 11-1
MEASURES OF RESOLUTION

DISCIPLINE	RESOLUTION METRIC
Optics	Rayleigh criterion Airy Disc diameter Blur diameter
Detectors	Number of detector elements
EO Systems (geometric approach)	Detector angular subtense Nyquist frequency Detector pitch
EO Systems (MTF approach)	Limiting resolution Effective-instantaneous-field-of-view
Monitors	TV limiting resolution Number of addressable pixels
Photo reconnaissance and remote sensing	Ground resolved distance

A single measure of spatial resolution cannot be satisfactorily used to compare all sensor systems. Spatial resolution does not provide information about total imaging capability or contrast sensitivity. The minimum resolvable contrast (MRC) simultaneously provides contrast sensitivity information and resolution information for monochrome systems. Resolution does not include the effects of system noise and is not related to sensitivity.

System resolution depends on optical diffraction, optical aberrations, detector pitch, detector size, digitization rate, electronic bandwidth, and monitor resolution. The most common measure of resolution used by the military is the detector angular subtense (DAS) because it is an easily understood metric for imaging systems. However, CCD performance is usually limited by the detector pitch. Note that different disciples use different metrics. The display designer does not "know" where the image originated. He only assumes that is "perfect" and that only the display degrades the image. As such, imaging system metrics are not routinely used by display designers.

When resolution is defined as the inverse of the system cutoff in angular space (typical for military applications), the back-of-the-envelope approximation provides the range at which the target can be "detected":

$$Range \approx \frac{target\ size}{resolution} \qquad (11\text{-}1)$$

This is the maximum range at which a periodic target can be faithfully reproduced.

11.1.1. ANALOG RESOLUTION METRICS

Analog measures of spatial resolution may be determined by: (1) the width of a point source image, (2) the minimum detectable separation of two point sources, (3) the spatial frequency at which the MTF drops to a defined level, or (4) the smallest detail resolved by an observer (Table 11-2). These measures assume that the system output image is a replica of the input scene (a linear, shift-invariant system).

Resolution may be defined from optical considerations. Diffraction of light through the optical aperture produces the smallest possible image spot size for an input point source. Diffraction measures include the Rayleigh criterion and the Airy disk diameter. The Airy disk is the bright center of the diffraction pattern produced by an ideal optical system. The Rayleigh criterion is a measure of the ability to distinguish two closely spaced objects when the objects are point sources. Optical aberrations and focus limitations increase the diffraction limited spot diameter to the blur diameter. Optical designers using ray tracing programs usually calculate the blur diameter. The blur diameter size is dependent on how it is specified (i.e., the fraction of light intensity encircled[8]).

Table 11-2
RESOLUTION MEASURES for ANALOG SYSTEMS

RESOLUTION	DESCRIPTION	TEST METHOD (usual units)
Rayleigh Criterion	Ability to distinguish two adjacent point sources	$\theta = 1220\,\lambda/D$ (mrad) (Calculated)
Airy Disk	Diffraction limited diameter produced by a point source	$\theta = 2440\,\lambda/D$ (mrad) (Calculated)
Blur Diameter	Actual minimum diameter produced by point source	Calculated from ray tracing (rad)
Limiting Resolution	Spatial frequency at which MTF = 0.02 to 0.10	Measured or calculated (cy/mm)
TV limiting resolution	Number of resolved lines per picture height	Measured with a wedged-shaped line pattern (TV lines per picture height)
Ground resolved distance	The smallest test target (1 cycle) that a photo interpreter can distinguish	Measured or calculated (feet or meters)
Ground resolution	An estimate of the limiting feature size seen by a photo interpreter	Measured (feet or meters)

The limiting resolution can be defined as that spatial frequency where the MTF drops to a fixed value, say, 2%, 5% or 10%. TV limiting resolution is determined from the finest detail that can be discerned by an observer when viewing stars, wedges or linear burst resolution patterns. TV limiting resolution is subjective and is roughly equated to when the MTF is approximately 3% (see Section 7.5.4., *MTF*, page 188). For commercial applications, the resolution is often specified by the spatial frequency where the MTF is 10%.

For aerial reconnaissance and associated image interpretation, resolution is measured by the ground resolved distance[9,10]. GRD is the minimum test target (one cycle) size that can be resolved on the ground by an experienced photo interpreter. The GRD is an objective measure of the physical ability of a camera to resolve standard contrast targets. The smallest detail that can be distinguished will have a physical width of GRD/2. The GRD is:

$$GRD = (Resolution) \cdot R_1 \qquad (11\text{-}2)$$

where R_1 is the slant range to the target. The reconnaissance community typically measures distance perpendicular to the line-of-sight. Therefore, Equation 11-2 does not contain a $\cos\theta$ correction factor for the target aspect angle. Any analog resolution measure listed in Table 11-2 or digital measure listed in Table 11-3 (see next section) can be used but the detector angular subtense and Nyquist frequency are the most common. GRD cannot be measured in the laboratory since it depends on the distance to the target, but, GRD may be calculated from an appropriate resolution measure.

Ground resolution is a subjective term that is a numerical estimate of the limiting features of objects to be examined. The system must be capable of resolving these features. For example, a system may require a ground resolution of four inches when examining the white center line of a highway. When examining granite boulders lying on sand beside the same highway, the system need only resolve, perhaps, two feet. The utility of the GRD is that a camera with the appropriate resolution can be selected for the task at hand. Although the GRD-based Imagery Interpretability Rating Scale was developed by the military, it can be modified for environmental remote sensing[11].

11.1.2. SAMPLED DATA SYSTEMS

New measures of resolution have been introduced with the advent of sampled data systems (Table 11-3). The detector angular subtense, DAS, is often used by the military to describe the resolution of systems when the detector is the limiting subsystem. If the detector element horizontal and vertical dimensions are different, then the DAS in the two directions is different. The instantaneous-field-of-view (IFOV) is the solid angle from which a single detector element senses radiation. It is a summary measure that includes both the optical and detector responses. If the optical blur diameter is small compared to the DAS, then the IFOV is approximately equal to the DAS. The IFOV is typically a measured quantity. Here, a point source transverses the detector element and the detector output is graphed as a function of angle. The IFOV is the full width one-half maximum amplitude of the resultant angular pulse shape. If the system output is measured, then the IFOV also depends on the electronic MTFs. The effective-instantaneous-field-of-view (EIFOV) offers an alternate measure of resolution (Figure 11-1). For many systems, the EIFOV and IFOV are approximately equal.

282 CCD ARRAYS, CAMERAS, and DISPLAYS

Table 11-3
MEASURES OF RESOLUTION for SAMPLED DATA SYSTEMS

RESOLUTION	DESCRIPTION	TEST METHOD (usual units)
Detector angular subtense	Angle subtended by one detector element	1000d/fl (mrad) (calculated)
Instantaneous-field-of-view	Angular region over which the detector senses radiation	Measured width at 50% amplitude (mrad)
Nyquist frequency	One-half of the sampling frequency	Calculated (cy/mm)
Detector pitch	Center-to-center spacing	Measured (mm)
Effective-instantaneous-field-of-view	One-half of the reciprocal of the object-space spatial frequency at which the MTF is 0.5	Measured or calculated (mrad)
Pixels	Number of detector elements or number of digital data points	A number (numeric)
Resels	Smallest region that contains unique information. A resel consists of one or more pixels.	Measured or calculated

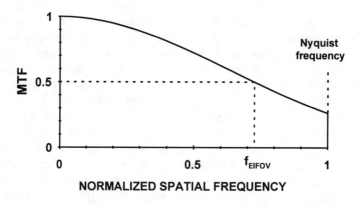

Figure 11-1. Definition of effective-instantaneous-field-of-view for an undersampled system. The spatial frequency has been normalized to the Nyquist frequency, f_N. EIFOV = $1/(2 f_{EIFOV})$.

If resolution is defined by the DAS, then a vanishingly small detector is desired. Although any detector size, optical aperture, and focal length can be chosen to select the resolution limit, the sensitivity is affected by the same parameters. As the resolution improves, the sensitivity degrades. Resolution and sensitivity factors are coupled into a single figure of merit: the minimum resolvable contrast or MRC (discussed in Chapter 12, *Minimum Resolvable Contrast*).

Sampled data systems may be limited (1) optically, (2) by the detector, or (3) by the array Nyquist frequency. For undersampled systems, the system MTF is defined only up to the Nyquist frequency (see Section 8.3.2., *Detector Array Nyquist Frequency*, page 209). If the detector pitch is d_{CC}, then the array Nyquist frequency is $f_N = 1/(2d_{CC})$. For CCDs the Nyquist frequency is the most often quoted number.

Digital processing is used for image enhancement and analysis. Because the pixels are numerical values in a regular array, mathematical transforms can be applied to the array. The transform can be applied to a single pixel, group of pixels, or the entire image. Many image processing designers think of images as an array of numbers that can be manipulated with little regard to who is the final interpreter.

In those cameras that have an analog output, the signal is digitized externally to the imaging system. The number of samples created by the external analog-to-digital converter (ADC) is simply a function of the ADC capability. Here, the sample number can be much greater than the number of detector elements. This higher number does not create more resolution. Any image processing algorithm that operates on this higher number must consider the sensor resolution.

An image may be partitioned into many discrete data points. Each point represents a location and spatially averaged local image intensity. Each point is considered a picture element or pixel. The number of pixels selected to represent the image depends on the number of detectors for a staring array or the sampling rate for a scanning array. Each point represents both the output of a detector element and an IFOV. Information is contained in a resolution element or resel. A resel is the smallest region in object space whose dimension is equal to the resolution in that dimension. In oversampled systems, the resel may consist of many pixels. For undersampled systems (e.g., CCD cameras), the resel typically equals the IFOV. The resolution measures provided in Tables 11-2 and 11-3 contain different ways of defining a resel.

Example 11-1
RESELS AND PIXELS

An imaging system consists of 512 x 512 elements. Assume that the fill-factor is 100% The camera's analog output is digitized at either 1024 samples/line or 2048 samples/line. What are the numbers of resels and pixels for the digitized video? For this example, the resel, IFOV, and DAS are equal.

With an 100% fill factor array, each pixel represents a resel. The external analog-to-digital converter creates either two pixels or four pixels for each resel. In both cases, the resel value remains the same. Increasing the sampling rate does not change the system's resolution but improves the repeatability of all measurements by reducing phasing effects that might occur in the analog-to-digital converter. Any image processing algorithm that operates on this higher number must consider the sensor resolution.

The highest spatial frequency that can be reproduced is limited by the array Nyquist frequency and not the pixel size or number of pixels.

Example 11-2
MONITOR PIXELS VERSUS SYSTEM RESELS

An imaging system is connected to a digital monitor that displays of 1024 x 1024 pixels. The staring array consists of 512 x 512 detectors. What is the system resolution?

Let each DAS be equal to a resel. Each resel is mapped onto four monitor pixels. In this case, the imaging system determines the system resolution, not the monitor. High-quality monitors only insure that the image quality is not degraded.

11.1.3. SHADE'S EQUIVALENT RESOLUTION

Shade[3] equated apparent image sharpness of a television picture to an equivalent pass band:

$$N_e = \int_0^\infty [MTF(f)]^2 \, df \qquad (11\text{-}3)$$

Sendall[12] modified Shade's equivalent resolution to

$$R_{eq} = \frac{1}{2N_e} \qquad (11\text{-}4)$$

With this modification, a detectors's R_{eq} is equal to the DAS. R_{eq} cannot be directly measured and is a mathematical construct used simply to express overall performance. As an approximation, the system resolution, $R_{eq\text{-}sys}$, may be estimated from the component equivalent resolutions, R_i, by

$$R_{eq\text{-}sys} \approx \sqrt{R_1^2 + R_2^2 + \ldots + R_n^2} \qquad (11\text{-}5)$$

Shade's approach using the square of the MTF emphasized those spatial frequencies at which the MTF is relatively high. It appears to be a good measure for classical systems in which the MTF is decreasing (such as a Gaussian distribution). The two MTFs shown in Figure 11-2 *could* have the same equivalent pass bands. Thus, a single number such as N_e should not be used to compare systems of different designs.

Shade's approach is appropriate for systems whose MTFs have similar shape. It probably should not be used (nor should any other image quality metric) when systems are compared with significantly different shapes. This, of course, includes system that can create significant aliasing. R_{eq} is usually different for the vertical and horizontal directions. The metric was developed to more accurately account for the human observer response. The applicability to machine vision systems has not been established at this time.

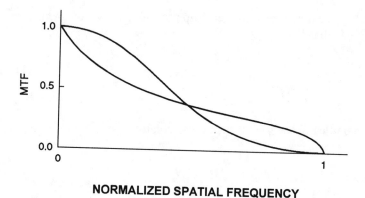

Figure 11-2. The MTFs of two different imaging systems. Both systems *could* have the same equivalent pass band.

11.2. MTFA

The area bounded by the system MTF and the eye's modulation threshold (M_t) is called the modulation transfer function area (MTFA) (Figure 11-3a). As the MTFA increases, the perceived image quality appears to increase. The spatial frequency at which the MTF and M_t intersect is a measure of the overall resolution. However, the resolution is not unique since M_t depends on the display viewing distance. The eye's inhibitory response (see Figure 10-24, page 267) is omitted from the MTFA approach. Mathematically,

$$MTFA = \int_0^{f_o} [MTF(f) - M_t(f)] df \qquad (11\text{-}6)$$

f_o is the limiting spatial frequency or intersection frequency. M_t is also called the aerial image modulation and demand modulation function. It is the minimum modulation required at a given spatial frequency to discern a difference.

According to Snyder[13], the MTFA appears to correspond well with performance in military detection tasks where the targets are embedded in noise. The noise elevates the eye's modulation threshold so that the MTFA decreases with increasing noise (Figure 11-3b).

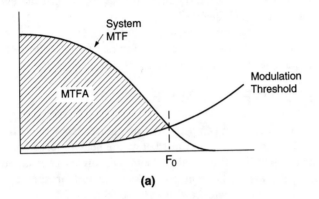

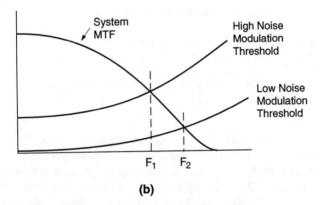

Figure 11-3. MTFA. The area between system MTF and eye's modulation threshold of the eye up to the intersection is a measure of image quality. The spatial frequency at the intersection is a measure of limiting resolution. The area and limiting resolution depend on the noise level within the imagery. (a) Low noise and (b) elevated noise.

288 CCD ARRAYS, CAMERAS, and DISPLAYS

For low-noise, general imagery, the eye's contrast sensitivity has little effect on the MTFA. The integrand in Equation 11-6 can be replaced with just the system MTF. Here, the high spatial frequencies are weighted too heavily compared to the low spatial frequencies and Shade's approach appears better.

From an analyst's viewpoint, the convolution theorem states that MTFs should be multiplied. The contrast sensitivity is proportional the eye's MTF. Therefore the system MTF should be multiplied by the contrast sensitivity or divided by the inverse (modulation threshold). However, the intersection frequency can be used as a measure of limiting resolution of the system. It can be used, for example, to calculate the GRD.

11.3. SUBJECTIVE QUALITY FACTOR

According to Granger and Cupery[2], the spatial frequency range important to image quality is in the region from just below the peak to three times the peak sensitivity of the eye. The Campbell-Robson high luminance data, and the eye models of Nill, Schultz, and deJong and Bakker support Granger and Cupery's approach whereas the Kornfeld-Lawson eye model does not (see Section 10.7., *Eye Response*, page 265). Recall that The Kornfeld-Lawson model is considered only applicable for spatial frequencies greater than the f_{PEAK}.

The normalized SQF in logarithmic units is:

$$SQF = K \int_{\log(3)}^{\log(12)} MTF_{SYS}(f_{eye}) \, d(\log(f_{eye})) \qquad (11\text{-}7)$$

K is a normalization constant. In the SQF approach, only those frequencies that are very important to the eye are included. The spatial frequency presented to the eye depends on the image size on the display, the distance to the display, and electronic zoom. Table 11-4 provides Granger and Cupery's interpretation of the SQF. These results are based on many observers viewing noiseless photographs with known MTF degradation.

The SQF is an adequate image quality metric for low-noise imagery when the illumination is fixed at a moderate level. As with the MTFA and Shade's equivalent pass band, the SQF is intended for general imagery only. It cannot be used for specific applications such as the legibility of alphanumeric characters.

Table 11-4
SUBJECTIVE QUALITY FACTOR

SQF	SUBJECTIVE IMAGE QUALITY
0.92	Excellent
0.80	Good
0.75	Acceptable
0.50	Unsatisfactory
0.25	Unusable

11.4. SQUARE-ROOT INTEGRAL

Barten[4-7] introduced the square-root integral (SQRI) as a measure of image quality. The SQRI is

$$SQRI = \frac{1}{\ln(2)} \int_{f_{min}}^{f_{max}} \sqrt{MTF_{SYS}(f_{eye}) \, CSF_{EYE}(f_{eye})} \, \frac{df_{eye}}{f_{eye}} \quad (11-8)$$

The multiplication overcomes the theoretical objection raised with the MTFA where the CSF was subtracted from the MTF. MTF_{SYS} includes all the subsystem MTFs up to and including the display. Barten modeled the contrast sensitivity function as

$$CSF(f_{eye}) = \frac{1}{M_t} = a f_{eye} e^{-bf_{eye}} \sqrt{1 + c e^{bf_{eye}}} \quad (11-9)$$

where

$$a = \frac{540 \left(1 + \frac{0.7}{L_{AVE}}\right)^{-0.2}}{1 + \frac{12}{w\left(1 + \frac{f_{eye}}{3}\right)^2}} \quad (11\text{-}10a)$$

$$b = 0.3\left(1 + \frac{100}{L_{AVE}}\right)^{0.15} \qquad (11\text{-}10b)$$

$$c = 0.06 \qquad (11\text{-}10c)$$

f_{eye} has units of cycles/degree, L_{AVE} is the average luminance in candelas per square meter, and w is the angular display size as seen by the observer. $1/\ln(2)$ allows SQRI to be expressed in just noticeable difference (JND) units. Both average luminance level and display size affect the CSF. This CSF is a good approximation over five orders of magnitude of luminance levels and for angular display sizes ranging from 0.5° to 60°.

This model includes the effects of various display parameters such as viewing ratio, luminance, resolution, contrast, addressability, and noise. Display viewing ratio and average luminance are incorporated in the eye's CSF (via w and L_{AVE} in the CSF equation). Display resolution (display MTF) is included in MTF_{SYS}.

Contrast loss due to reflected ambient lighting effectively reduces the display MTF. If ΔL is the additional luminance due to reflected ambient light, then the modulation depth is reduced by the factor:

$$\eta_{DISPLAY} = \frac{L_{AVE}}{L_{AVE} + \Delta L} \qquad (12\text{-}11)$$

This factor is the same for all spatial frequencies of interest and the perceived effect on image quality is incorporated into the SQRI by multiply MTF_{SYS} by $\sqrt{\eta_{DISPLAY}}$.

The display parameters dictate the limits of integration. f_{MIN} is that spatial frequency that provides one-half cycle across the display. f_{MAX} is the maximum spatial frequency of the display and this is related to display addressability (see Section 7.6., *Addressability*, page 189). For example, if the number of active lines is N_{ACTIVE}, then

$$f_{MAX} = \frac{(Kell\,factor)\,N_{ACTIVE}}{2}\frac{R}{H} \qquad (12\text{-}12)$$

The small angle approximation was used to calculate field-of-view subtended by the observer. It is the display height, H, divided by the viewing distance, R.

Noise increases the contrast sensitivity function. Since the SQRI must decrease with noise, Barten added noise as an additional factor to the eye MTF:

$$MTF_{t-noise}(f_{eye}) = \sqrt{MTF_{eye}^2(f_{eye}) + (k_{noise}M_n)^2} \qquad (12\text{-}13)$$

M_n can be calculated from the noise power spectral density presented to the eye[5]. Barten predicted the quality of imagery when both static (pattern noise) and dynamic (temporally varying) noise are present. He considered both one-dimensional and two-dimensional noise sources. His approach encompasses actual sensor operation where different noise components may exist in the horizontal and vertical directions. As such, he considered some of the three-dimensional noise components used for military applications (discussed in Section 12.2., *Three-Dimensional Noise Model*).

11.5. REFERENCES

1. H. Kusaka, "Consideration of Vision and Picture Quality - Psychological Effects Induced by Picture Sharpness," in *Human Vision, Visual Processing and Digital Display*, B. E. Rogowitz, ed., SPIE Proceedings Vol. 1077, pp. 50-55, (1989).
2. E. M. Granger and K. N. Cupery, "An Optical Merit Function (SQF) Which Correlates With Subjective Image Judgments," *Photographic Science and Engineering*, Vol. 16, pp. 221-230 (1972).
3. O. H. Shade, Sr., "Image Gradation, Graininess, and Sharpness in Television and Motion Picture Systems," published in four parts in *SMPTE Journal*: "Part I: Image Structure and Transfer Characteristics," Vol. 56(2), pp. 137-171 (1951); "Part II: The Grain Structure of Motion Pictures - An Analysis of Deviations and Fluctuations of the Sample Number," Vol. 58(2), pp. 181-222 (1952); "Part III: The Grain Structure of Television Images," Vol. 61(2), pp. 97-164 (1953); "Part IV: Image Analysis in Photographic and Television Systems," Vol. 64(11), pp. 593-617 (1955).
4. P. G. Barten, "Evaluation of Subjective Image Quality with the Square-root Integral Method," *Journal of the Optical Society of America. A*, Vol. 17(10), pp. 2024-2031 (1990).
5. P. G. Barten, "Evaluation of the Effect of Noise on Subjective Image Quality," in *Human Vision, Visual Processing and Digital Display II*, J. P. Allenbach, M. H. Brill, and B. E. Rogowitz, eds., SPIE Proceedings Vol. 1453, pp. 2-15 (1991).
6. P. G. Barten, "Physical Model for the Contrast Sensitivity of the Human Eye," in *Human Vision, Visual Processing and Digital Display III*, B. E. Rogowitz, SPIE Proceedings Vol. 1666, pp. 57-72 (1992).
7. P. G. Barten presents short courses at numerous symposia. See, for example, "Display Image Quality Evaluation," *Application Seminar Notes*, SID International Symposium held in Orlando, Fl (May 1995). Published by the Society for Information Display, Santa Ana, CA, or "MTF, CSF, and SQRI for Image Quality," IS&T/SPIE's Symposium on Electronic Imaging: Science & Technology, San Jose Ca (February 1995).

8. L. M. Beyer, S. H. Cobb, and L. C. Clune, "Ensquared Power for Obscured Circular Pupils With Off-Center Imaging," *Applied Optics*, Vol. 30(25), pp. 3569-3574 (1991).
9. Air Standardization Agreement: "Minimum Ground Object Sizes for Imaging Interpretation," Air Standardization Co-ordinating Committee report AIR STD 101/11 (31 December 1976).
10. Air Standardization Agreement: "Imagery Interpretability Rating Scale," Air Standardization Co-ordinating Committee report AIR STD 101/11 (10 July 1978).
11. J. D. Greer and J. Caylor, "Development of an Environmental Image Interpretability Rating Scale," in *Airborne Reconnaissance XVI*, T. W. Augustyn and P. A. Henkel, eds., SPIE Proceedings Vol. 1763, pp. 151-157 (1992).
12. J. M. Lloyd, *Thermal Imaging*, pg. 109, Plenum Press, New York (1975).
13. H. L. Synder, "Image Quality and Observer Performance," in *Perception of Displayed Information*, L. M. Biberman, ed., pp. 87-118, Plenum Press, New York, NY (1973).

12
MINIMUM RESOLVABLE CONTRAST

Most image quality metrics were developed for high contrast imagery. Here, image quality is closely linked with the MTF. High MTF cameras provide better image quality and therefore provide better resolution under optimum conditions. While most resolution metrics are applicable to high contrast targets, the relative rating of systems may change for low contrast scenes. A high MTF system with excessive noise will not permit detection of low contrast targets.

The military is interested in detecting and recognizing targets at the greatest possible range. This is achieved by increasing the system gain until noise is apparent and the imagery is "snowy." Since the distinction between target and background is measured by contrast, the signal threshold is expressed as the minimum resolvable contrast (MRC). MRC applies to monochrome systems only. It is a measure of an observer's ability to detect a bar target embedded in noise. The Johnson criterion relates real targets to equivalent bar chart targets. Thus, the link between real targets and the MRC is the spatial frequency of the equivalent bar chart target.

By providing an angular resolution, it is simple to determine the range: R = (target size)/(resolution). Therefore, for MRC calculations, object-space spatial frequency (cy/mrad) is used. CCD performance was specified in the image-space spatial frequency domain. All the MTFs in Chapter 10 were functions on f_i. The conversion is $f_{os} = fl\, f_i$ where fl is the effective optical focal length.

In the 1950s, Shade[1] modeled the resolution of photographic film and TV sensors as a function of light level. His approach is the frame work of all models used today. In the 1970s, Rosell and Willson[2] applied Shade's results to thermal imaging systems and low-light-level televisions. Since then, most modeling efforts have concentrated on infrared imaging systems. However, the same modeling equations apply to all imaging systems. Different forms exist to accommodate the terminology associated with different technologies (scanning versus staring, visible versus thermal imaging systems, etc.).

MRC was originally developed for vidicons and low-light-level televisions. The approach taken here extends the thermal imaging system performance model (popularly called FLIR92) methodology[3] to visible systems. FLIR92 calculates the minimum resolvable temperature (MRT) where the target-background radiance difference is characterized by a differential temperature. FLIR92 methodology and the three-dimensional noise model[4] are reformatted for CCD cameras. The 3-D noise model was developed to describe the various noise sources present in thermal imaging systems. These same noises exist in lesser amounts within CCD cameras.

The MRC approach is applied to visible systems since it relates how the sensor performs using terminology that characterizes the human visual system. The observer does not see the same contrast at the display as that presented by the target to the camera. Both the displayed contrast and brightness can be adjusted on the display so the observed contrast may be significantly different from what is present at the entrance aperture of the imaging system. Observer threshold is expressed as the perceived signal-to-noise ratio rather than the perceived contrast.

Within the real world, there is a probability associated with every parameter. The target contrast is not one number but a range of values that follow a diurnal cycle. The atmospheric transmittance is not fixed but can change in minutes. There appears to be an overwhelming set of combinations and permutations. Therefore, only a few representative target contrasts and a few representative atmospheric conditions are selected and the performance range is calculated for these conditions.

The Nyquist frequency limits the highest frequency that can be reproduced. The specific manner that the Nyquist frequency is included in the model greatly affects range performance. This is a result of the assumptions used and is not necessarily representative of actual system performance. While no model is perfect, they are still excellent for comparative analyses.

The symbols used in this book are summarized in the *Symbol List* (page xiv) and it appears after the *Table of Contents*.

12.1. THE OBSERVER

The original Night Vision Laboratory static performance model[5] and FLIR92 use the Kornfeld-Lawson eye model (see Section 10.7., *Eye Response*, page 265). These models were validated by the US Army when detecting tank-sized targets at modest ranges. Insufficient data exists to say with certainty which of the eye models available is best. Because of this and other model uncertainties, range predictions cannot be placed on an absolute scale. *All analyses, no matter which eye model is used, must be used only for comparative performance purposes.*

12.1.1. PERCEIVED SIGNAL-TO-NOISE RATIO

The eye/brain "filters" enhance the displayed SNR. The perceived signal-to-noise ratio, referred back to the array output, is

$$SNR_P = k \frac{MTF_{SYS} \Delta n_{pe}}{\langle n_{SYS} \rangle} \frac{1}{(eye\ spatial\ filter)(eye\ temporal\ filter)} \quad (12\text{-}1)$$

MTF_{SYS} includes both the camera MTFs (see Chapter 10, *System MTF*, page 233) and the eye MTF. All calculations are performed with object-space units (e.g., cy/mrad). Δn_{pe} is the difference in the number of photo electrons generated by the target and its immediate background. k is a constant that depends on the optical aperture diameter, focal length, detector spectral quantum efficiency, etc. (see Equation 6-2, page 156). $\langle n_{SYS} \rangle$ is the total system noise.

This approach differs from the camera signal-to-noise ratio presented in Section 6.1.2. (page 160). There, the target was considered very large so the system MTF did not affect the signal strength. The rationale for using the MTF here, is that observer metrics (i.e., MRC) are based on the visibility of high frequency bar targets.

Although the eye's modulation threshold is reported as a J-shaped curve (see Figure 10-24, page 267), the actual shape depends on the noise power spectral density[6-8]. If the noise is restricted to certain spatial frequencies, then the detection of targets of comparable spatial frequencies becomes more difficult. The observer's ability to see a specific spatial frequency target depends on the noise content in the neighborhood of that spatial frequency (Figure 12-1). Low spatial frequency noise components will interfere with detecting low frequency targets (large objects). Mid-spatial frequency noise increases the modulation threshold curve at mid-frequencies and so on. These noise factors are included in the MRC model via the three-dimensional noise model. As such, they are not included in the eye MTF.

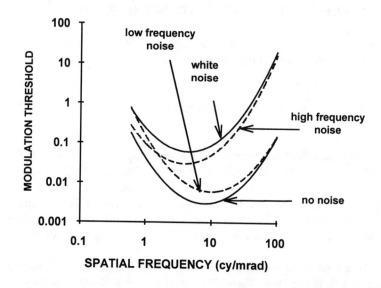

Figure 12-1. Effects of spectral noise on the eye's modulation threshold. The MRC should approximately follow the shape of the modulation threshold curve.

For nearly all imaging systems, square waves (bar targets) are used for system characterization. Using the CTF_{SYS} (See Section 9.4., *Contrast Transfer Function*, page 228), the perceived signal-to-noise ratio is

$$SNR_P = k' \frac{CTF_{SYS} \Delta n_{pe}}{\langle n_{SYS} \rangle (eye\ spatial\ filter)(eye\ temporal\ filter)} \quad (12\text{-}2)$$

k' is a constant that includes k and the conversion from MTF to CTF.

Selecting a threshold value that is required to just perceive a target and inverting Equation 12-2 provides a minimum detectable target to background difference:

$$\Delta n_{MINIMUM} = SNR_{TH} \frac{\langle n_{SYS} \rangle (eye\ spatial\ filter)(eye\ temporal\ filter)}{k'\ CTF_{SYS}} \quad (12\text{-}3)$$

The system noise may consist of many components and the eye may integrate each component differently. β_i is the "filter" that interprets noise source $\langle n_i \rangle$:

$$\Delta n_{MINIMUM} = SNR_{TH} \frac{\sqrt{\langle n_1^2 \rangle \beta_1 + ... + \langle n_m^2 \rangle \beta_m}}{k' \, CTF_{SYS}} \qquad (12\text{-}4)$$

or

$$\Delta n_{MINIMUM} = SNR_{TH} \frac{\langle n_1 \rangle}{k' \, CTF_{SYS}} \sqrt{\beta_1 + ... + \frac{\langle n_m^2 \rangle}{\langle n_1^2 \rangle} \beta_m} \qquad (12\text{-}5)$$

A test bar-target is approximated by a square wave of infinite extent. If the system is band-limited, then only the fundamental of the square wave has sufficient amplitude to contribute to the response (valid for high frequencies). The fundamental's amplitude is $4/\pi$ times the square wave amplitude. The eye is sensitive to the average value of the first harmonic and the average value of a half-cycle sine wave is $2/\pi$. Therefore, the conversion from a square wave (CTF_{SYS}) to a sinusoid (MTF_{SYS}) requires a factor of $8/\pi^2$ when viewed by an observer:

$$\Delta n_{MINIMUM} = SNR_{TH} \frac{\pi^2}{8} \frac{\langle n_1 \rangle}{k \, MTF_{SYS}} \sqrt{\beta_1^2 + ... + \frac{\langle n_m^2 \rangle}{\langle n_1^2 \rangle} \beta_m^2} \qquad (12\text{-}6)$$

Following FLIR92 terminology[3], $\langle n_1 \rangle$ is the system shot noise. The other noise terms represent all other noise sources. The three-dimensional noise model quantifies each $\langle n_i^2 \rangle$.

In Section 6.1.2., *Camera Signal-to-Noise Ratio* (page 164) a target SNR was calculated. The background light level was assumed to be zero so that the photon shot noise was associated with the target only. For target detection, the target exists on a positive background. Both the target and the background regions provide photons that the camera converts to photo electrons. $n_{pe\text{-}T}$ and $n_{pe\text{-}B}$ are the photoelectrons associated with the target and background respectively. There is shot noise associated with both. The "average" shot noise variance is

$$\langle n_1^2 \rangle = \langle n_{SHOT}^2 \rangle \approx \frac{n_{pe-T} + n_{pe-B}}{2} \qquad (12\text{-}7)$$

A variety of definitions exist for contrast. The contrast, as defined in the original MRC derivation, is the target luminance minus the background luminance divided by the background luminance. Since both the target and background are assumed to be illuminated by the same source (sun, moon, etc.), the contrast depends on reflectances.

$$C_O = \frac{|L_T - L_B|}{L_B} = \frac{|\rho_T - \rho_B|}{\rho_B} \qquad (12\text{-}8)$$

Historically, the target was always considered darker than the background. However, the target can be brighter than the background and there does not appear to be any difference between the detectability of objects that are of negative or positive contrast.

Substituting into Equations 12-7 and 12-8 into Equation 12-6, and calling the minimum detectable contrast the MRC, provides

$$MRC = SNR_{TH} \frac{\pi^2}{8} \frac{1}{k\, MTF_{SYS}} \sqrt{\frac{MRC + 2}{2 n_{pe-B}}} \sqrt{\beta_1 + \ldots + \frac{\langle n_m^2 \rangle}{\langle n_{SHOT}^2 \rangle} \beta_m} \qquad (12\text{-}9)$$

Since MRC appears on both sides of the equation, it can be solved either iteratively or as a quadratic equation. The MRC is a family of curves dependent on the background level (Figure 12-2). MRC approaches a maximum value of unity as the system MTF approaches zero. Thus MRC is linked to MTF but the resolution (i.e., the spatial frequency where the contrast approaches one) depends on the background.

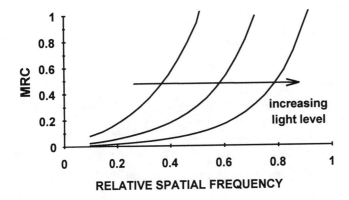

Figure 12-2. MRC is a family of curves that depend on the background light level.

12.1.2. PERCEIVED RESOLUTION

Perceived resolution depends on the size of the target, ambient lighting, target-to-background contrast, and noise. These issues were not considered when discussing display resolution since they are relatively constant.

For signals embedded in noise, the situation is different. The eye is then very sensitive to the signal-to-noise ratio. Slight variations affect target visibility. As illustrated in Figure 12-3, the signal increases faster than the accompanying photon shot noise. Given a threshold of perceptibility, as the SNR increases, more detail can be seen because more signals rise above the threshold. Thus, camera "resolution" depends on background light level. This "resolution" is strictly due to the eye's response and is not related to hardware performance. In contrast, hardware resolution is typically specified as that spatial frequency where the hardware MTF is, say 10% and is measured under relatively high signal-to-noise ratio conditions.

Figure 12-4 illustrates the perceived resolution as a function of light level. The apparent resolution increases with light level. Since the SNR is proportional to contrast, the "resolution" is also a function of contrast. High contrast targets produce higher resolution. Again, this is strictly due to the eye's response and is not related to hardware limitations.

300 CCD ARRAYS, CAMERAS, and DISPLAYS

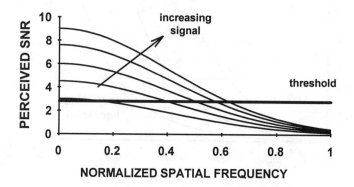

Figure 12-3. As the target signal increases, the SNR increases. For a fixed threshold, increasing the signal appears to increase the resolution.

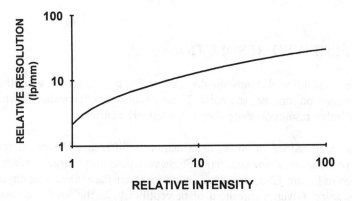

Figure 12-4. Perceived resolution as a function of light level.

12.1.3. THE EYE/BRAIN "FILTER"

The eye/brain system is probably the most difficult system to model. Although many models exist, the two most often employed are the matched filter and the synchronous integrator model. With the matched filter, the eye acts like a tuned circuit. Here, the perceived SNR is affected only by spatial frequencies that are near the target's spatial frequency. All other spatial frequencies are ignored. This is not a filter in the usual sense since the signal and noise are "filtered" by it. Rather, visual psychophysical data suggest that the eye act as if it were a filter that can be described mathematically by a filter function. With the synchronous integrator model, the eye integrates over an angular region defined by the target edges.

Scott and D'Agostino state[9] *The most frequently encountered methods...describe the eye/brain spatial integration using either a matched filter or a synchronous integrator model .... they yield virtually identical .... predictions ... Any potential difference ... will be lost in the inherent error of the .... measurements. For these reasons, and because ... a synchronous integrator is somewhat simpler ....,* FLIR92 has been written as a synchronous integrator model. Russell and Willson[2] used the synchronous integrator model for television systems and their approach is presented in this chapter.

12.2. THREE-DIMENSIONAL NOISE MODEL

The three-dimensional noise model[4] provides the basic framework for analyzing the various noise sources. It was developed to describe the noise in thermal imaging systems. However, the methodology can be applied to CCD cameras.

The noise is divided into a set of eight components that relate temporal and spatial noise to a three-dimensional coordinate system (Figure 12-5). This approach allows full characterization of all noise sources including random noise, pattern noise, streaks, rain, 1/f noise, and any other artifact. Analyzing the noise in this manner has the advantage of simplifying the understanding of complex phenomena by breaking it down into a manageable set of components. The method simplifies the incorporation of complex noise factors into model formulations.

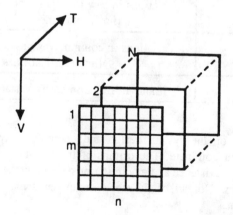

Figure 12-5. Three-dimensional noise model coordinate system illustrating data set N_{TVH}.

The T-dimension is the temporal dimension representing the video framing sequence. The other two dimensions (horizontal and vertical) provide spatial information. For a staring array, n and m indicate detector element locations in the horizontal and vertical directions respectively.

Table 12-1 describes the three-dimensional noise components. The subscripts describe the noise "direction." σ_{TVH} represents rms noise value of three-dimensional data set m x n x N. σ_{VH} is the rms noise value after averaging in the T-direction. Its data set contains m x n elements. And so on.

Table 12-1
THREE-DIMENSIONAL NOISE DESCRIPTORS

3-D NOISE COMPONENT	DESCRIPTION
σ_{TVH}	Random 3-D noise
σ_{VH}	Spatial noise that does not change from frame-to-frame
σ_{TH}	Variations in column averages that change from frame-to-frame (rain)
σ_{TV}	Variations in row averages that change from frame-to-frame (streaking)
σ_{V}	Variations in row averages that are fixed in time (horizontal lines or bands)
σ_{H}	Variations in column averages that are fixed in time (vertical lines)
σ_{T}	Frame-to-frame intensity variations (flicker)

Depending on the system design and operation, any one of these noise components can dominate. The origins of these components are significantly different and the existence and manifestation depend on the specific design of the imaging system. Not all of the noise components may be present in every imaging system.

Assuming the noise components are independent, the total system noise is

$$\sigma_{SYS} = \sqrt{\sigma^2_{TVH} + \sigma^2_{TH} + \sigma^2_{TV} + \sigma^2_{VH} + \sigma^2_{H} + \sigma^2_{V} + \sigma^2_{T}} \qquad (12\text{-}10)$$

Independence was a fundamental assumption during the development of the three-dimensional noise model. The global average, S, is the average intensity level and has no variation for a fixed data set. For nearly all systems, σ_T is considered negligible compared to σ_{TVH} and therefore is also omitted from the data set. σ_{TVH} consists of the shot noise and the noise floor. Pattern noise is incorporated though σ_{VH}, σ_V, and σ_H. Currently, only σ_{TVH} is predicted and the remaining noise components must be determined from measurements or estimates. Figures 12-6 through 12-8, generated by the SIA software[10], illustrate the effect of these noise sources. Figure 12-6 is the basic image. In Figure 12-7, FPN, PRNU, and random noise appear similar in a single frame. Random noise changes from frame-to-frame whereas pattern noise does not. Figure 12-8 illustrates how σ_{TV} or σ_V affects the visibility of horizontal bars. Scanning systems often exhibit this type of noise. Pixels will always have different responsivities. When the row-average responsivities are different in TDI sensors, banding occurs[11].

Figure 12-6. Ideal image with $\sigma_{SYS} = 0$. Generated by SIA software[10].

304 CCD ARRAYS, CAMERAS, and DISPLAYS

Figure 12-7. Image with random, FPN, or PRNU noise.

Figure 12-8. Image with dominant horizontal banding (high σ_{TV} or σ_V).

While the three-dimensional noise model was created to describe noise sources in thermal imaging systems, many of the same noises exist in all imaging systems. The FLIR92 default values should apply to all imaging systems (including CCD cameras) built to similar designs.

With all systems, the most prevalent noise source is assumed to be temporal noise (σ_{TVH}). For staring arrays, the next most important noise source is σ_{VH}. Table 12-2 lists the FLIR92 recommended[3] values for PtSi staring arrays operating in the 3 - 5 μm. Other systems using different detectors may perform differently. These values were obtained empirically.

For scanning systems, temporal row noise (σ_{TV}), and fixed row noise (σ_V) are the important contributors. Table 12-3 lists the FLIR92 default values appropriate for scanning systems with HgCdTe detectors operating in the 8 - 12 μm region. These values apply to all linear arrays where the image motion is perpendicular to the array axis.

Table 12-2
DEFAULT VALUES
for STARING SYSTEMS

RELATIVE NOISE	DEFAULT VALUE
σ_{VH}/σ_{TVH}	0.40
σ_{TV}/σ_{TVH}	0
σ_V/σ_{TVH}	0
σ_{TH}/σ_{TVH}	0
σ_H/σ_{TVH}	0

Table 12-3
DEFAULT VALUES
for SCANNING SYSTEMS

RELATIVE NOISE	DEFAULT VALUES
σ_{VH}/σ_{TVH}	0
σ_{TV}/σ_{TVH}	0.25
σ_{V}/σ_{TVH}	0.25
σ_{TH}/σ_{TVH}	0
σ_{H}/σ_{TVH}	0

12.3. THE MRC MODEL

The MRC is a measure of an observer's ability to detect a signal embedded in noise. The MRC model is a so-called *static* model in that the target is stationary and no search is required (or at least the observer knows where to look). The observer has an unlimited amount of time for target discrimination. As with most models, the system is assumed to be linear and shift invariant.

This section applies the FLIR92 methodology to CCD camera performance. The fundamental difference between the MRC and MRT is the method by which the target is characterized. In the infrared spectral region, the target-to-background differential signal is specified by a temperature difference. In the visible spectral region, the differential signal is specified by the contrast. The MRT derivation can be found in the literature[5,12-15].

Following FLIR92 methodology, the horizontal and vertical MRCs are

$$MRC_H(f_x) = SNR_{TH} \frac{\pi^2}{8} \frac{1}{k\,MTF_H(f_x)} \sqrt{\frac{MRC+2}{2n_{pe-B}}} K_H(f_x) \quad (12\text{-}11)$$

and

$$MRC_V(f_y) = SNR_{TH} \frac{\pi^2}{8} \frac{1}{k\,MTF_V(f_y)} \sqrt{\frac{MRC+2}{2n_{pe-B}}} K_V(f_y) \quad (12\text{-}12)$$

$K_H(f_x)$ and $K_V(f_y)$ are summary noise factors that combine the three-dimensional noise model components with the eye's spatial and temporal integration capabilities. $MTF_H(f_x)$ is the product of all the horizontal MTFs and $MTF_V(f_y)$ is the product of all the vertical MTFs (see Chapter 10, *System MTF*, page 233). In the original development of the MRC model only eye/brain spatial and temporal integration factors were considered. Later models included the eye MTF.

The summary noise factors are

$$K_H(f) = \left(E_T E_V E_H + \frac{\sigma_{VH}^2}{\sigma_{TVH}^2} E_V E_H + \frac{\sigma_{TH}^2}{\sigma_{TVH}^2} E_T E_H + \frac{\sigma_H^2}{\sigma_{TVH}^2} E_H \right)^{\frac{1}{2}} \quad (12\text{-}13)$$

$$K_V(f) = \left(E_T E_V E_H + \frac{\sigma_{VH}^2}{\sigma_{TVH}^2} E_V E_H + \frac{\sigma_{TV}^2}{\sigma_{TVH}^2} E_T E_V + \frac{\sigma_V^2}{\sigma_{TVH}^2} E_V \right)^{\frac{1}{2}} \quad (12\text{-}14)$$

For staring systems, the default (see Table 12-2) MRCs become

$$MRC_H(f_x) = \frac{\pi^2}{8} \frac{SNR_{TH}}{k\,MTF_H(f_x)} \sqrt{\frac{MRC+2}{2n_B}} \sqrt{E_V E_H \left(E_T + \frac{4}{10} \right)} \quad (12\text{-}15)$$

$$MRC_V(f_y) = \frac{\pi^2}{8} \frac{SNR_{TH}}{k\,MTF_V(f_y)} \sqrt{\frac{MRC+2}{2n_B}} \sqrt{E_V E_H \left(E_T + \frac{4}{10} \right)} \quad (12\text{-}16)$$

For scanning systems (applies to linear arrays) the default (see Table 12-3) MRCs are

$$MRC_H(f_x) = \frac{\pi^2}{8} \frac{SNR_{TH}}{k\,MTF_H(f_x)} \sqrt{\frac{MRC+2}{2n_B}} \sqrt{E_T E_V E_H} \quad (12\text{-}17)$$

$$MRC_V(f_y) = \frac{\pi^2 SNR_{TH}}{8\,k\,MTF_V(f_y)} \sqrt{\frac{MRC+2}{2n_B}} \sqrt{E_V \left(E_T E_H + \frac{1}{4} E_T + \frac{1}{4} \right)} \quad (12\text{-}18)$$

E_T, E_H, and E_V respectively represent the temporal, horizontal spatial and vertical spatial integration afforded by the eye/brain interpreter. The model assumes that the eye temporally integrates perfectly and continuously over the eye integration time. The noise is considered uncorrelated from frame-to-frame so that it adds in quadrature. When the information update rate, F_R, is high enough, and if there is signal correlation between adjacent frames:

$$E_T = \frac{1}{F_R t_e} \qquad (12\text{-}19)$$

A 5:1 aspect ratio 3-bar target ($T_{ASPECT} = 5$) is used to measure the MRC, whereas a 7:1 aspect ratio 4-bar target is typically used for MRT measurements. The bars are oriented vertically for horizontal measurements and horizontally for vertical measurements. Thus, the equations used depend on the bar orientation. A vertical and horizontal MRC exists for each bar orientation. Table 12-4 lists the nomenclature. Before the two-dimensional MRC approach, only $E_{H\text{-}H}$ was used for horizontal MRC measurements. Similarly, only $E_{V\text{-}V}$ was used for vertical MRC measurements. The additional components, $E_{V\text{-}H}$ and $E_{H\text{-}V}$, are used in the two-dimensional MRC (discussed in Section 12.5., *Two-dimensional MRC*).

Table 12-4
MRC EQUATION NOMENCLATURE

MEASUREMENT	BAR ORIENTATION	MODEL NOMENCLATURE
Horizontal MRC	Vertical	Horizontal MRC: $E_{H\text{-}H}$ Vertical MRC: $E_{V\text{-}H}$
Vertical MRC	Horizontal	Horizontal MRC: $E_{H\text{-}V}$ Vertical MRC: $E_{V\text{-}V}$

The noise filters and noise power spectrum were described in Section 6.1.2., *Camera Signal-to-Noise Ratio* (page 160). The noise spectrum is modified by the electronic filters that occur after the insertion of noise. Since noise is assumed to originate in the detector, MTF_{OPTICS}, MTF_{MOTION}, and $MTF_{DETECTOR}$ are not part of noise filters. As appropriate, the remaining MTFs provided in Chapter 10, *System MTF* (page 233), are components of the noise filters. For equation brevity, the horizontal and vertical noise filters are listed as $H_{NF\text{-}H}(f)$ and $H_{NF\text{-}V}(f)$ respectively. Since electrical signals are typically read out in a serial data stream (consistent with video standard timing), the noise appears in the horizontal direction only.

For horizontal MRC, the bars are oriented vertically and the bar aspect ratio, T_{ASPECT}, appears in the vertical summary noise factors. For a bar target whose fundamental spatial frequency is f_o, the summary noise factors are

$$E_{H-H}(f_o) = S_{CCH} \int_0^\infty S(f) |H_{NF-H}(f)|^2 \, sinc^2\left(\frac{f}{2f_o}\right) df \quad (12\text{-}20)$$

$$E_{V-H}(f_o) = S_{CCV} \int_0^\infty |H_{NF-V}(f)|^2 \, sinc^2\left(\frac{T_{ASPECT}f}{2f_o}\right) df \quad (12\text{-}21)$$

S_{CCH} and S_{CCV} are the angular detector center-to-center spacings (detector pitch) in the horizontal and vertical directions respectively expressed in mrad. f is the electrical frequency expressed in object frequency units. For electrical signals in the video chain (expressed in Hertz),

$$f_x = \frac{t_{LINE}}{HFOV} f_{elec} \quad \frac{cy}{mrad} \quad (12\text{-}22)$$

t_{LINE} is the active line time (see Table 5-1, page 128) and HFOV it the horizontal field-of-view.

For the vertical MRC, the bar aspect ratio is in the horizontal summary noise factor:

$$E_{H-V}(f_o) = C_{SSH} \int_0^\infty S(f) |H_{NF-H}(f)|^2 \, sinc^2\left(\frac{T_{ASPECT}f}{2f_o}\right) df \quad (12\text{-}23)$$

$$E_{V-V}(f_o) = C_{SSV} \int_0^\infty |H_{NF-V}(f)|^2 \, sinc^2\left(\frac{f}{2f_o}\right) df \quad (12\text{-}24)$$

In the early literature, only the horizontal MRC was calculated and only random temporal noise was modeled. With no other noise sources,

$$MRC_H(f_x) = \frac{\pi^2}{8} \frac{SNR_{TH}}{k \, MTF_H(f_x)} \sqrt{\frac{MRC+2}{2n_{pe-B}}} \sqrt{E_V E_H E_T} \quad (12\text{-}25)$$

310 CCD ARRAYS, CAMERAS, and DISPLAYS

E_T, E_H and E_V are given by Equations 12-19, 12-20, and 12-21 respectively.

If MTFs are near unity and the noise is white [S(f) ≈ 1] over the spatial frequencies of interest (i.e., when f < 2f_c), then

$$E_H = E_{H-H}(f_o) \approx S_{CCH} \frac{f_o}{T_{ASPECT}} \quad (12\text{-}26)$$

$$E_V = E_{V-H}(f_o) \approx S_{CCV} f_o \quad (12\text{-}27)$$

Then

$$MRC(f_o) = \frac{\pi^2}{8} \frac{SNR_{TH}}{k\,MTF(f_c)} \sqrt{\frac{MRC+2}{2n_{pe-B}}} \sqrt{\frac{1}{F_R t_e}} \sqrt{\frac{S_{CCH} S_{CCV}}{T_{ASPECT}} f_o^2} \quad (12\text{-}28)$$

The original SNR formulation[16] contained an $(a/A)^{1/2}$ factor where "a" is the target area at the image plane and "A" is the effective area of the photosensitive device. The MRC is the inverse of the SNR and therefore should contain $(A/a)^{1/2}$. This is identical to the last square root term in Equation 12-28. "A" is the photosensitive area that contributes to noise. With vidicons, "A" represented the entire image surface. With CCD arrays, it is the effective pixel size, $S_{CCH}S_{CCV}$. The target size is $(1/f_o)(T_{ASPECT}/f_o)$. Equation 12-28 reduces to the basic MRC equation used by most researchers[17,18] prior to the introduction of the three-dimensional noise model.

12.4. SNR_{TH} and t_e

The FLIR92 documentation[19] states: *SNR_{TH} and the eye integration time, t_e, are most often the model parameters used to "tune" an MRTD prediction to a set of measurements.* These values are treated as constants that are adjusted so that laboratory data matches predicted values.

The 1975 NVL model documentation[20] states: .... *unfortunately, universal values for these constants do not exist. The values recommended at this time are SNR_{TH} = 2.25 and t_e = .2.*

FLIR92 documentation continues[19] *The luminances associated with about a 0.1 second eye integration time agree well with the display experiments conducted in 1988 ... (darkened room and optimal viewing).... In higher ambient light levels conditions, such as may be encountered with fielded systems, greater display luminances can be expected and thus a faster eye integration time may be appropriate... NVESD recommends setting SNR_{TH} equal to 2.5.... Though psychophysical data shows that SNR_{TH} is a function of target frequency, a value of 2.5 represents a reasonable average.*

t_e is somewhat nebulous and its value depends on the specific task on hand and ambient light level. FLIR92 recommends changing t_e according to the available light level (Figure 12-9).

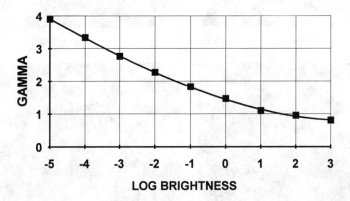

Figure 12-9. Eye integration values as a function of light level. FLIR92 documentation recommends using $t_e = 0.1$ sec for laboratory MRT predictions. The same technique can be applied to MRC.

While many arguments exist for selecting the "correct" values, SNR_{TH} and t_e are simply constants. Both SNR_{TH} and t_e have a range of acceptable values. No matter what value is selected for t_e, consistency throughout the model must be maintained. t_e is also used in the linear motion MTF (see Section 10.4.1., *Linear Motion*, page 251).

A word of clarification is necessary here. SNR_{TH} is approximately the signal-to-noise value that the eye/brain perceives in the image at the detection threshold. It includes, in part, both temporal and spatial integration effects. It is *not* the signal-to-noise ratio of the video signal nor the SNR of the luminance on the display. Depending on the noise characteristics and the target characteristics, the eye can detect[21] signals whose video SNR is as low as 0.05 (Figure 12-10). The eye/brain perceives this SNR to be about 2.25.

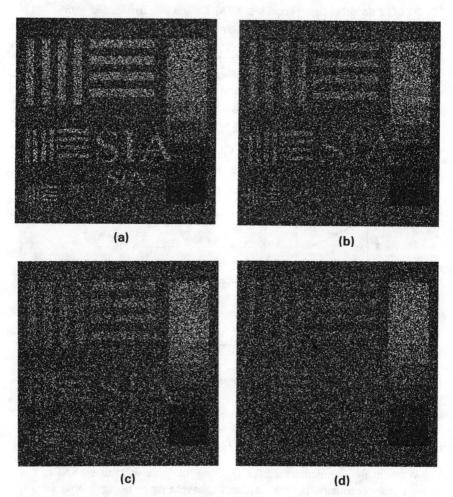

Figure 12-10. Variations in the signal-to-noise ratio, (a) SNR = 2, (b) SNR = 1, (c) SNR = 0.75, and (d) SNR = 0.5. The imagery was computer generated with the SNRs listed. The tonal transfer characteristics of the laser printer (to create the hard copy) and halftone (to print the image) may have affected the final SNR. The actual SNRs in this printed copy have not been measured. These images should be considered as representative of the ability to perceive low SNR imagery.

The ability to perceive targets embedded in noise depends on the target spatial extent with respect to the noise spatial frequencies. Large bars are easier to perceive compared to small targets that "disappear" into the noise. Figure 12-10 only illustrates the eye's spatial integration capabilities. With live video, the random noise is changing from frame to frame and the eye's temporal integration provides an additional enhancement of $F_R t_e$. For EIA 170 timing, this is about a factor of three to six.

12.5. TWO-DIMENSIONAL MRC

Although FLIR92 is called two-*dimensional*, it is a two-*directional* model. That is, the threshold is predicted along two orthogonal axes, taken as the vertical and horizontal directions. With unequal MTFs, the predicted MRC will be different in the two directions. The two directional values are brought together with a two-dimensional spatial frequency that approximates the response of an equivalent radially symmetric system.

As with the MRTs, the two-dimensional MRC is created by taking the geometric average of the horizontal and vertical MRC frequencies at each contrast value (Figure 12-11):

$$f_{2D} = \sqrt{f_V f_H} = \sqrt{f_x f_y} \qquad (12\text{-}29)$$

In this manner, the two-dimensional MRC is forced to asymptotically approach the mean values of the vertical and horizontal cutoff frequencies. The two-dimensional MRC is a mathematical construct that is used for range performance predictions. It cannot be measured since f_{2D} does not exist.

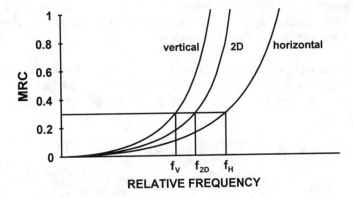

Figure 12-11. The two-dimensional MRC is mathematically created from the horizontal and vertical MRCs. It is really a two-*directional* MRC.

314 CCD ARRAYS, CAMERAS, and DISPLAYS

12.6. RANGE PREDICTIONS

Performance modeling assumes that the target is a single spatial frequency. This is an approximation since real targets consist of a complex superposition of spatial frequencies. Targets are actually a continuum of spatial frequencies from zero on up. The system MTF limits the target spatial frequencies that can be displayed. This chapter provides an overview to the range prediction methodology. A detailed discussion of system range performance can be found in reference 22.

The Johnson criterion provides the link between the target angular subtense and the spatial frequency scale on the MRC graph. The target's inherent contrast, C_O, is modified by the atmospheric conditions (attenuation and path radiance). When the target contrast is reduced to the MRC curve, the performance range is obtained. Depending on the Johnson criterion selected, this intersection is the detection, recognition, or identification range.

12.6.1. CONTRAST TRANSMITTANCE

Intuitively, it seems that any phenomena that prevent us from visually perceiving an object will affect all imagery the same way. During a hazy day, light scattered into the eye (path radiance) reduces the visual contrast. Here, objects that are far away appear as a neutral white and we are unable to distinguish any features. While the eye acts in this manner, imaging systems do not. Imaging systems simply respond to radiance differences.

However, real systems have automatic gain circuitry that may be activated by the path radiance signal. If so, the effect is to reduce system gain. Path radiance will partially fill the charge wells.

For most applications, the target and background are illuminated by the same source (sun, moon, etc.). The apparent contrast, C_R, is the inherent contrast modified by the atmosphere. Dropping the wavelength notation for equation brevity,

$$C_O = \frac{\Delta L}{L} = \frac{e^{-\sigma_{ATM}R}[\rho_T - \rho_B]L_e}{e^{-\sigma_{ATM}R}\rho_B L_e + L_{ATM}} = C_o \frac{1}{1 + \frac{L_{ATM}}{\rho_b L_e}e^{\sigma_{ATM}R}} \quad (12\text{-}30)$$

L_e is the ambient illumination that irradiates the target and background. It is the sky, sun, or moon modified by the intervening atmospheric transmittance. σ_{ATM} is the atmospheric scattering cross section and $\exp(-\sigma_{ATM}R)$ is the atmospheric transmittance. The path radiance, L_{ATM}, is due to light scattered into the line-of-sight and the total amount received is an integral over the path length[23]. When only scattering is present (a reasonable approximation for the visible spectral band), $L_{ATM} \approx (1 - \exp(-\sigma_{ATM}R))L_{SKY}$. L_{SKY} depends on the viewing direction and the location of the sun. $\rho_B L_e$ is the background luminance. For convenience, the $L_{SKY}/\rho_B L_e$ is called the sky-to-ground ratio (SGR):

$$C_R = C_O \frac{1}{1 + SGR(e^{\sigma_{ATM}R} - 1)} \qquad (12\text{-}31)$$

The SGR is approximately $0.2/\rho_B$ for a clear day ($L_{SKY}/L_e \approx 0.2$) and $1/\rho_B$ for an overcast day ($L_{SKY}/L_e \approx 1$). When the SGR is one, the received contrast is simply the inherent contrast reduced by the atmospheric transmittance:

$$C_R \approx C_O e^{-\sigma_{ATM}R} \qquad (12\text{-}32)$$

Meteorological range is defined quantitatively by the Koschmieder formula[24]:

$$R_{VIS} = \frac{1}{\sigma_{ATM}} \ln\left(\frac{1}{C_{TH}}\right) \qquad (12\text{-}33)$$

C_{TH} is the modulation threshold at which 50% of the observers would see the target. The use of the scattering cross section in this definition rather than the extinction coefficient implies that absorption of particles at visual wavelengths is small enough to ignore. This view is probably justified except in cases of polluted air. Koschmieder set C_{TH} to 0.02 and evaluated σ at $\lambda = 0.555$ μm. Then

$$Transmittance = T_{ATM} = e^{-\sigma_{ATM}R} = e^{-\frac{3.912}{R_{VIS}}R} \qquad (12\text{-}34)$$

Table 12-5 lists σ_{ATM} as a function of standard weather conditions. However, σ_{ATM} decreases with increasing wavelength. For visibilities less than 6 km, σ_{ATM} can be approximated by

$$\sigma_{ATM}(\lambda) \approx \left(\frac{3.912}{R_{VIS}}\right)\left(\frac{0.55}{\lambda}\right)^{0.585(R_{VIS})^{1/3}} \quad (12\text{-}35)$$

For much greater visibilities,

$$\sigma_{ATM}(\lambda) \approx \left(\frac{3.912}{R_{VIS}}\right)\left(\frac{0.55}{\lambda}\right)^{0.5(R_{VIS})^{1/3}} \quad (12\text{-}36)$$

Table 12-5
INTERNATIONAL VISIBILITY CODE

DESIGNATION	VISIBILITY	SCATTERING COEFFICIENT, σ_{ATM}
Dense fog	0 - 50 meters	> 78.2 km^{-1}
Thick fog	50 - 200 m	19.6 - 78.2 km^{-1}
Moderate fog	200 - 500 m	7.82 - 19.6 km^{-1}
Light fog	500 - 1 km	3.92 - 7.82 km^{-1}
Thin fog	1 - 2 km	1.96 - 3.92 km^{-1}
Haze	2 - 4 km	0.978 - 1.96 km^{-1}
Light haze	4 - 10 km	0.391 - 0.978 km^{-1}
Clear	10 - 20 km	0.196 - 0.391 km^{-1}
Very clear	20 - 50 km	0.0782 - 0.196 km^{-1}
Exceptionally clear	> 50 km	< 0.782 km^{-1}

12.6.2. JOHNSON CRITERIA

The Johnson[25] methodology has become known as the equivalent bar pattern approach. Observers viewed military targets through image intensifiers and were asked to detect, decide the orientation, recognize and identify the targets. Air Force tri-bar charts whose bars had the same contrast as the scaled models were also viewed and the maximum resolvable bar pattern frequency was determined. The number of bars per critical object dimension was increased until the bars could just be individually resolved. In this way, detectability was correlated with the sensor's threshold bar pattern resolution (Table 12-6). These results became the foundation for the discrimination methodology used today. *Critical* dimension refers to the target dimension that must be discerned for detection, recognition, or identification. In the early literature, the minimum dimension was called the critical dimension.

Table 12-6
JOHNSON'S RESULTS
(From reference 25)

DISCRIMINATION LEVEL	MEANING	CYCLES ACROSS CRITICAL TARGET DIMENSION
Detection	An object is present (object versus noise).	1.0 ± 0.025
Orientation	The object is approximately symmetrical or unsymmetrical and its orientation may be discerned (side view versus front view).	1.4 ± 0.35
Recognition	The class to which the object belongs (e.g., tank, truck, man).	4.0 ± 0.80
Identification	The object is discerned with sufficient clarity to specify the type (e.g., T-52 tank, friendly jeep).	6.4 ± 1.50

Although Johnson used 6.4 cycles across the critical target dimension for identification, studies with thermal imaging systems at the Night Vision Laboratory suggested that eight cycles are more appropriate for identification. Orientation is a less popular discrimination level. Because today's standards are based on Johnson's work, they are labeled as the Johnson criteria though they are not the precise values recommended by him.

318 CCD ARRAYS, CAMERAS, and DISPLAYS

The original Johnson methodology was validated using common module thermal imaging systems. Those systems typically had a 2:1 difference in horizontal to vertical resolution. The vertical resolution was limited by the detector pitch Nyquist frequency. For validation purposes, military vehicles were used and the critical dimension tended to be in the vertical direction although the predictions were based on horizontal performance. Converting to a two-dimension model requires removing the directional bias imposed by the one-dimensional model. This is achieved by reducing all values by 25%, or equivalently, multiplying[26] all discrimination values by 0.75. This adjustment then provides the same range performance predictions whether predicted by the one-dimensional or two-dimensional model. Table 12-7 provides the commonly used discrimination levels. By convention, each 50% probability is labeled as N_{50} (e.g., in one-dimension, N_{50} is 1, 4, or 8 cycles for detection, recognition, or identification respectively).

Table 12-7
COMMON DISCRIMINATION LEVELS

TASK	DESCRIPTION	One-dimension N_{50}	Two-dimension N_{50}
Detection	The blob has a reasonable probability of being an object being sought.	1.0	0.75
Classical recognition	Object discerned with sufficient clarity that its specific class can be differentiated.	4.0	3.0
Identification	Object discerned with sufficient clarity to specify the type within the class.	8.0	6.0

12.6.3. TARGET TRANSFER PROBABILITY FUNCTION

The results of threshold experiments, such as Johnson's, provide an approximate measure of the 50% probability of discrimination level. Results of several field tests[20] provided the cumulative probability of discrimination or target transfer probability function (TTPF) (Table 12-8). The TTPF can be used for all discrimination tasks by simply multiplying the 50% probability of performing the task by the TTPF multiplier. For example, from Table 12-7, the probability of 95% recognition is $2N_{50} = 2(4) = 8$ cycles across the target's critical dimension.

Table 12-8
DISCRIMINATION CUMULATIVE PROBABILITY

PROBABILITY OF DISCRIMINATION	MULTIPLIER
1.00	3.0
0.95	2.0
0.80	1.5
0.50	1.0
0.30	0.75
0.10	0.50
0.02	0.25
0	0

An empirical fit[27] to the data provides

$$P(N) = \frac{\left(\dfrac{N}{N_{50}}\right)^E}{1 + \left(\dfrac{N}{N_{50}}\right)^E} \qquad (12\text{-}37)$$

where

$$E = 2.7 + 0.7\left(\frac{N}{N_{50}}\right) \qquad (12\text{-}38)$$

Figure 12-12 illustrates the full TTPF for the three levels of discrimination as identified by Johnson. Because of the variability in the population, when some people can only detect the target, others will recognize it and still a smaller portion will identify the target. It is this variation that leads to the wide variations seen in field test results.

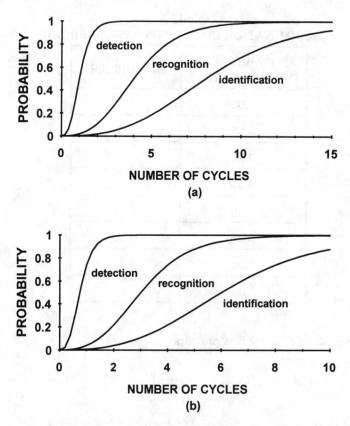

Figure 12-12. Target transfer probability function for detection, recognition, and identification. (a) one-dimensional, and (b) two-dimensional. The curves overlap so that when only a small portion of the population can identify the target, a larger portion will recognize the target. An even larger portion will detect the target.

12.6.4. RANGE PREDICTION METHODOLOGY

The MRC abscissa is converted into a range scale using a target discrimination value. With the two-dimensional model, the critical dimension, h_c, is the square root of the target area, A_{TARGET}. For rectangular targets, $h_c = (HW)^{1/2}$ where H and W are the target's height and width respectively. The number of equivalent cycles across the target is

$$N_T = \frac{\sqrt{A}}{R} f_{2D} = \frac{h_c}{R} f_{2D} \qquad (12\text{-}39)$$

Consider, for example, for a truly rectangular target. If the rectangular target size is 2 x 8 meters and six cycles are required across the critical dimension, then the conversion from spatial frequency to range is

$$R = \frac{\sqrt{2 \cdot 8}}{6} f_{2D} = \frac{4}{6} f_{2D} \qquad (12\text{-}40)$$

Six cycles represent 95% probability of recognition for the two-dimensional model. The intersection of the MRC curve and the target apparent contrast, C_R, versus range curve is the range at which the target can be discerned according to the discrimination level selected (Figure 12-13).

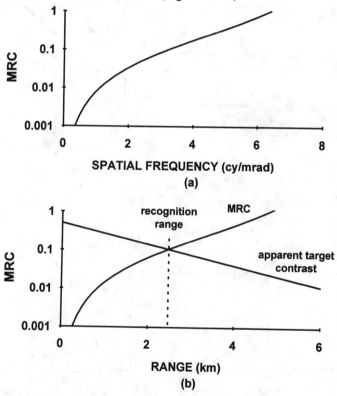

Figure 12-13. Graphical method to obtain the recognition range. (a) Two-dimensional MRC and (b) conversion of spatial frequency into range. $N_{50} = 3$, $h_c = 4$ m, and $C_0 = 0.5$. C_R is a straight line when plotted in semi-logarithmic coordinates. The visibility is 6.2 km. The two-dimensional 50% probability recognition range is approximately 2.5 km.

322 CCD ARRAYS, CAMERAS, and DISPLAYS

The target transfer probability function is used to predict the probability of range performance. Here, a range is selected and the C_R is calculated. This value intersects the MRC curve at what is called the critical frequency. Multiplying the angular subtense by the critical frequency provides the number of cycles across the target. Referring to the TTPF (see Figure 12-12), the probability is determined for that particular range. Then a new range is selected and the process is repeated until the entire probability function is determined (Figure 12-14).

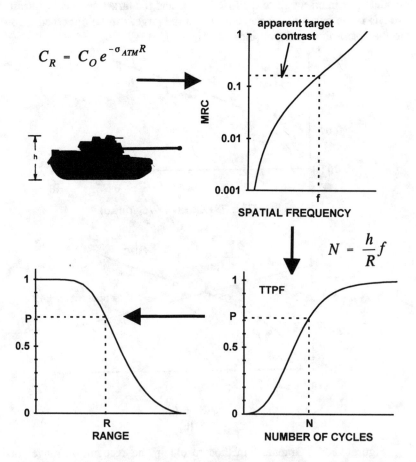

Figure 12-14. Methodology to determine range performance probability.

12.6.5. SAMPLING EFFECTS

In Chapter 8, *Sampling Theory*, page 199, it was stated that the highest frequency that could be faithfully reproduced is the Nyquist frequency. CCD cameras are undersampled systems and are limited to a spatial frequency dictated by the detector center-to-center spacing. In the development of the MRC equation, sampling effects were ignored.

Since no frequency can exist above the Nyquist frequency, many researchers represent the MTF as zero above the Nyquist frequency. This forces the MRC to approach one at the Nyquist frequency (Figure 12-15). When the probability of detection is calculated using the target transfer probability function, the Nyquist limit reduces the range (Figure 12-16).

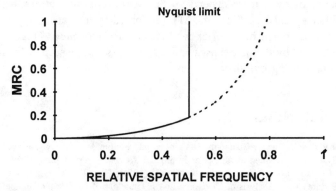

Figure 12-15. Representative MRC for a staring array sensor. At Nyquist frequency, the MRC is assumed to be one. The abrupt increase at Nyquist frequency is used by some researchers.

This range reduction is an artifact of the model. It is real when trying to discriminate sinusoids or bar targets. Real targets, however, are composed of many spatial frequencies ranging from zero and on up. Target features that have spatial frequencies above Nyquist will be distorted. No data exist on the detection or recognition of distorted imagery and it is, therefore, unwise at this juncture to extend the MRC model past the Nyquist frequency. The analyst must recognize that the Nyquist frequency limit may have an overly severe impact on target discrimination range. The specific manner in which the Nyquist frequency is included in the model has a dramatic effect on range performance. This is a result of the assumptions used and is not necessarily representative of actual system performance.

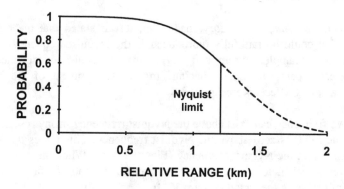

Figure 12-16. Probability of detection versus range for a staring array. The shapes of the curves depend on the MRC, contrast, atmospheric transmittance, and N_{50}. Since the MTF is forced to zero at the Nyquist frequency, the range hits a brick wall at Nyquist frequency. This is an artifact created by the model assumptions. Targets can be seen at longer ranges but aliasing will distort edges.

12.7. REFERENCES

1. O. H. Shade, Sr., "Image Gradation, Graininess, and Sharpness in Television and Motion Picture Systems," published in four parts in *SMPTE Journal*: "Part I: Image Structure and Transfer Characteristics," Vol. 56(2), pp. 137-171 (1951); "Part II: The Grain Structure of Motion Pictures - An Analysis of Deviations and Fluctuations of the Sample Number," Vol. 58(2), pp. 181-222 (1952); "Part III: The Grain Structure of Television Images," Vol. 61(2), pp. 97-164 (1953); "Part IV: Image Analysis in Photographic and Television Systems," Vol. 64(11), pp. 593-617 (1952)
2. F. A. Rosell and R. H. Willson, "Performance Synthesis of Electro-Optical Sensors," Air Force Avionics Laboratory Report AFAL-TR-72-229, Wright Patterson AFB, OH (1972).
3. *FLIR92 Thermal Imaging Systems Performance Model, User's Guide*, NVESD document UG5008993, Fort Belvoir, VA (1993).
4. J. D'Agostino and C. Webb, "3-D Analysis Framework and Measurement Methodology for Imaging System Noise," in *Infrared Imaging Systems: Design, Analysis, Modeling and Testing II*, G. C. Holst, ed., SPIE Proceedings Vol. 1488, pp. 110-121 (1991).
5. W. R. Lawson and J. A. Ratches, "The Night Vision Laboratory Static Performance Model Based on the Matched Filter Concept," in *The Fundamentals of Thermal Imaging Systems*, F. Rosell and G. Harvey, eds., NRL Report 8311, pp. 159-179, Naval Research Laboratory, Washington, D.C. (1979).
6. S. Daly, "Application of a Noise Adaptive Contrast Sensitivity Function in Image Data Compression," *Optical Engineering*, Vol. 29(8), pp. 977-987 (1990).
7. H. Pollehn and H. Roehrig, "Effect of Noise on the Modulation Transfer Function of the Visual Channel," *Journal of the Optical Society of America*, Vol. 60, pp. 842-848 (1970).
8. A. Van Meeteren and J. M. Valeton, "Effects of Pictorial Noise Interfering With Visual Detection," *Journal of the Optical Society of America A*, Vol. 5(3), pp. 438-444 (1988).

9. L. Scott and J. D'Agostino, "NVEOD FLIR92 Thermal Imaging Systems Performance Model," in *Infrared Imaging Systems: Design, Analysis, Modeling and Testing III*, G. C. Holst, ed., SPIE Proceedings Vol. 1689, pp. 194-203 (1992).
10. SIA (System Image Analysis) software is available from JCD Publishing, 2932 Cove Trail, Winter Park, FL 32789.
11. T. S. Lomheim and L. S. Kalman, "Analytical Modeling and Digital Simulation of Scanning Charge-Coupled Device Imaging Systems," in *Electro-Optical Displays*, M. A. Karim, ed., pp. 551-560, Marcel Dekker, New York (1992).
12. J. A. Ratches, W. R. Lawson, L. P. Obert, R. J. Bergemann, T. W. Cassidy, and J. M. Swenson, "Night Vision Laboratory Static Performance Model for Thermal Viewing Systems," ECOM Report ECOM-7043, pg. 34, Fort Monmouth, NJ (1975).
13. J. M. Lloyd, *Thermal Imaging Systems*, pp. 182-194, Plenum Press, NY (1975).
14. F. A. Rosell, "Laboratory Performance Model," in *The Fundamentals of Thermal Imaging Systems*, F. Rosell and G. Harvey, eds., NRL Report 8311, pp. 85-95, Naval Research Laboratory, Washington, D.C. (1979).
15. R. L. Sendall and F. A. Rosell, "Static Performance Model Based on the Perfect Synchronous Integrator Model," in *The Fundamentals of Thermal Imaging Systems*, F. Rosell and G. Harvey, eds., NRL Report 8311, pp. 181-230, Naval Research Laboratory, Washington, D.C. (1979).
16. F. A. Rosell, "Video, Display, and Perceived Signal-to-Noise Ratios," in *The Fundamentals of Thermal Imaging Systems*, F. Rosell and G. Harvey, eds., NRL Report # 8311, pp. 49-83, Naval Research Laboratory, Washington, D.C. (1979).
17. W. W. Frame, "Minimum Resolvable and Minimum Detectable Contrast Prediction for Vidicon Cameras," *SMPTE Journal*, Vol. 104(1), pp. 21-26 (1985).
18. W. W. Frame, "Minimum Resolvable and Minimum Detectable Contrast Prediction for Monochrome Solid-State Imagers," *SMPTE Journal*, Vol. 96(5), pp. 454-459 (1987).
19. *FLIR92 Thermal Imaging Systems Performance Model, Analyst's Reference Guide*, NVESD document RG5008993, pp. ARG-12 - 13, Fort Belvoir, VA (1993).
20. J. A. Ratches, W. R. Lawson, L. P. Obert, R. J. Bergemann, T. W. Cassidy, and J. M. Swenson, "Night Vision Laboratory Static Performance Model for Thermal Viewing Systems," ECOM Report ECOM-7043, pg. 56, Fort Monmouth, NJ (1975).
21. F. A. Rosell, "Psychophysical Experimentation," in *The Fundamentals of Thermal Imaging Systems*, F. Rosell and G. Harvey, eds., NRL Report # 8311, pg. 225, Naval Research Laboratory, Washington, D.C. (1979).
22. G. C. Holst, *Electro-Optical Imaging System Performance*, Chapter 19, 20, and 21, JCD Publishing, Winter Park, FL (1995).
23. L. Levi, *Applied Optics*, pp. 118-124, Wiley and Sons (1980).
24. W. E. K. Middleton, *Vision Through the Atmosphere*, University of Toronto Press (1958).
25. J. Johnson, "Analysis of Imaging Forming Systems," in *Proceedings of the Image Intensifier Symposium*, pp. 249-273, Warfare Electrical Engineering Dept., US Army Engineering Research and Development Laboratories, Ft. Belvoir, VA (1958). This article is reprinted in *Selected Papers on Infrared Design*, R. B. Johnson and W. L. Wolfe, eds., SPIE Proceedings Vol. 513, pp. 761-781 (1985).
26. L. Scott and R. Tomkinson, "An Update on the C^2NVEO FLIR90 and ACQUIRE Sensor Performance Model," in *Infrared Imaging Systems: Design, Analysis, Modeling and Testing II*, G. C. Holst, ed., SPIE Proceedings Vol. 1488, pp. 99-109 (1991).
27. J. D. Howe, "Electro-Optical Imaging System Performance Prediction," in *Electro-Optical Systems Design, Analysis, and Testing*, M. C. Dudzik, ed., pg. 92. This is Volume 4 of the *Infrared and Electro-Optical Systems Handbook*, J. S. Accetta and D. L. Shumaker, eds., copublished by Environmental Research Institute of Michigan, Ann Arbor, MI and SPIE Press, Bellingham, WA (1993).

APPENDIX

EFFECTIVE FOCAL LENGTH

Most optical systems are composed of many individual lenses and mirrors. Each element may have a different refractive index and shape to minimize the system aberrations. The optical design can be considered as a single thick element (Figure A-1). For modeling purposes, it is treated as a single thin element with an effective focal length (Figure 2-6, page 23). The clear aperture is not necessarily the diameter of an optical element. The clear aperture limits the amount of scene radiation reaching the detector and is determined by the optical design.

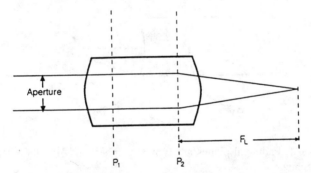

Figure A-1. The optical system can be considered as a single thick lens. P_1 and P_2 are the principal surfaces. The effective focal length is measured from the second principal surface. The clear aperture limits the amount of scene radiation reaching the detector. Although shown as planes, the principal surfaces are typically spherical.

f-number

The radiometric equations were derived from plane geometry and paraxial ray approximations. When using solid angles, the image incidance is proportional to $\sin^2(U')$ where U' is the maximum angle subtended by the lens. For paraxial rays, the principal surface is considered a plane. This representation is shown in most text books.

Lens design theory[1] assumes that the principal surface is spherical - every point on the surface is exactly a focal-length distance away (Figure A-2). This "sine corrected" lens has zero spherical aberration. Thus $\sin^2(U')$ is equal to $1/(4F^2)$. The numerical aperture is another measure of the energy collected by the optical system. When the image is in air (refractive index of unity) the numerical aperture is

$$NA = \sin U' = \frac{1}{2F} \quad (A\text{-}1)$$

Since the largest angle is $\pi/2$, the smallest theoretical value for F is ½. This theoretical limit on F is not obvious from the radiometric equation.

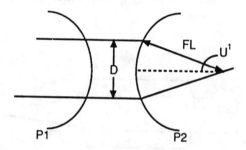

Figure A-2. Principal planes are typically spherical. $\sin(U') = D/2fl = 1/2F$ (in air).

Not all optical systems have spherical principal planes. Often we assume the principal plane is flat. For hand-drawn optical layouts, the lenses are approximated by thin lenses and paraxial rays are used (large f-numbers). Here, $\sin(\theta) \approx \theta$ and Equation A-1 reduces to F = fl/D. Then $\sin^2(U')$ is $1/(4F^2 + 1)$. Whether $1/(4F^2 + 1)$ or $1/4F^2$ is used in the radiometric equations depends upon the f-number definition. If the analyst calculates the f-number from the effective lens diameter and focal length, then $1/(4F^2 + 1)$ should be used. If the optical designer provides the f-number, then the analyst must consult with him to insure that the appropriate factor is used. For large f-numbers, the factors are approximately equal.

REFERENCE

1. W. J. Smith, *Modern Optical Engineering*, second edition, pp. 142-145, McGraw-Hill, New York (1990).

INDEX

Absorption coefficient 92, 242
Addressability 189
Adjacent pixel requirement 190
Airy disk 279
Aliasing 202
    display 218
    frame grabber 217
Alternating pixel requirement 190
Amplifier noise 110
Analog-to-digital converter 148
Anti-aliasing filter 204, 248
Anti-blooming drain 78
Anti-parallel clock 45
Aperture correction 145, 261
Array
    frame interline transfer 61
    frame transfer 58
    full frame transfer 55
    interline transfer 60
    linear 54
    TDI 66
Atmospheric transmittance 315
Averaging filter 258

Back side illuminated 95
Beat frequency 212
Bilinear readout 54
Binning 53, 258
Birefringent 73, 140, 205, 248
Blackbody 19
Blanking 127
Blur diameter 279
Boost filter 145, 261
Broadcast standard
    NTSC 126
    PAL 126
    SECAM 126
Butterworth filter 260

Camera formula 22
Camera operation 123
Camera performance 155
Cathode ray tube 173
Channel block 44
Character recognition 195
Charge aggregation 53

Charge conversion 82
Charge grouping 53
Charge transfer efficiency 79, 244
Chicken wire 152
CIE 32
Collimator 235
Color correction 138
Color filter array 70, 209, 248
Color temperature 19, 34, 176
    display 33
Component analog video 131
Composite video 131
Contrast 197
Contrast transfer function 229
Contrast transmittance 314

Convolution 225
Correlated double sampling 83
Cutoff
    detector 208
    optical 239

Dark current 73
Dark noise 73, 108
Dark pixel 76
Defects 87
Depletion region 43, 92
Depletion width 242
Detector angular subtense 281
Detector cutoff 208
Detector pitch 209
Differential nonlinearity 148
Diffraction limited 239
Diffusion 82, 241
Diffusion length 44, 93
Digital filter 256
Digital recorder 9
Digital television 134
Dirac delta 224
Discrimination methodology 317
Display 169
    aliasing 218
    color 176
    color temperature 33
    flat panel 170
    monochrome 175
    spot size 178

330 CCD ARRAYS, CAMERAS, and DISPLAYS

DISTORTION 212
Drain 78
Dynamic range 103, 158

Effective-instantaneous-field-of-view 281
EIA 170 126
EIA 170A 126
EIA 343A 130
Equivalent background illumination 167
Equivalent bar pattern 317
Excess noise 165
Eye
    integration time 310
    modulation threshold 266, 295
    synchronous integrator 300
    tuned circuit 300

F-number 24, 327
False resolution 252
Fiber-optic bundle 163
Fiber-optic coupling 152
Fictitious spatial frequency 233
Field integration 62
Fill factor 60, 209, 240
Filter
    anti-alias 204, 248
    averaging 258
    boost 261
    butterworth 260
    digital 256
    optical anti-alias 248
    post-reconstruction 260
Fixed pattern noise 113
Flat field 180
Flicker 173
FLIR92 293
Floating diffusion 82
Floating diode 82
Frame grabber 217
Frame integration 62
Frame interline transfer 61
Frame transfer 58
Full-frame transfer 55

Gamma 142, 175
Gate 42
Grand Alliance 136
Ground resolution 281
Ground resolved distance 280

High Definition Television 135, 178

Illuminant
    A, B, C, D 32
ILLUMINATION LEVEL
    artificial 31
    maximum 158
    minimum 158
    natural 30
Image distortion 212
Image intensifier 149, 269
Image quality 11, 275
    square-root integral 289
    subjective quality factor 288
Image space 235
Imagery Interpretability Rating Scale 281
In-phase 214, 270
Instantaneous-field-of-view 281
Integral nonlinearity 148
Intensified CCD 150, 163
Interline transfer 60
IRE units 132
Isolation pixel 78

Jaggies 200
Jitter 251
Johnson criteria 317
Johnson noise 109

Kell factor 130, 186, 271
Knee 138
Koschmieder 315
KTC noise 109

Lambertian source 18, 28
Line spread function 227
Line-pair 236
Linear array 54
Linear filter theory 224
Linear-phase-shift 219
Low light television 149
Lumogen 95

Matched filter 300
Matrixing 72, 132
Mean-variance 115
Meteorological range 315
Microchannel plate 149, 164
Microlens 85, 211
Military 7

INDEX 331

Minification 165
Minimum resolvable contrast
    see MRC
Minimum resolvable temperature 293
Modulation threshold 266
Modulation transfer function
    see MTF
Modulation transfer function area 286
Moiré pattern 182, 199
Monitor 169
MOS capacitor 42
Motion 251
MRC 293
    two-dimensional 313
MRT 293
MTF 219
    average 271
    boost 261
    butterworth 260
    charge transfer efficiency 244
    detector 208, 240
    diffusion 241
    digital filter 256
    display 262
    eye 265
    Gaussian 252
    image intensifier 269
    linear motion 251
    median 271
    non-limiting 269
    optical 238
    optical anti-alias 248
    polychromatic 239
    post-reconstruction 260
    random motion 252
    sample and hold 259
    sample-scene 270
    separable 234
    TDI 245
Multi-pinned phasing 74

Noise
    1/f 110
    ADC 111
    amplifier 110
    excess 165
    fixed pattern 113
    Johnson 109
    kTC 109
    photoresponse

    nonuniformity 113
    quantization 111
    reset 109
    shot 108
    summary factors 307
    three-dimensional 301
    white 110
Noise equivalent bandwidth 109
Noise equivalent exposure 102
Noise equivalent input 161
Noise equivalent reflectance difference 162
Normalization 35
NTSC 125
Nyquist frequency 199, 277, 323

Object space 235
Observer 171
Optical anti-alias filter 248
Organic phosphor 95
Out-of-phase 214, 270
Output structure 82
Overflow drain 78
Oversampling 200

PAL 126
Path radiance 314
Phase
    four 45
    three 49
    two 50
    virtual 52
Phase reversal 252
Phase transfer function 219
Phosphor 95, 176
Photo gate 42
Photometry 26
Photon standard 90
Photon transfer 90, 115
Photopic 26
Photoresponse nonuniformity 113
Pitch 209
    triad 176
Pixel 181, 283
Pixel density 184
Planck's blackbody law 19
Point spread function 226
Poisson statistics 108
Polychromatic 239
Polysilicon 93
Post-reconstruction filter 260

Progressive scan 64
Pseudo-interlacing 62

Quantization noise 111
Quantum efficiency 92

Radiative transfer 17
Radiometry 16
Random walk 44
Range prediction 314, 320
Raster 172
Rayleigh criterion 279
Reflectance 298
Relay lens 163
Resel 283
Reset noise 109
Resolution 186, 189, 277, 299
    false 252
    spurious 252
    TV limiting 187
Resolution/addressability ratio 190, 263
Responsivity 37, 97
RS 170 126
RS 170A 126

Sample-and-hold 259
Sample-scene 270
Sampling 199, 323
Sampling theorem 202
Saturation equivalent exposure 102
Scotopic 26
SECAM 126
Setup 133
Shade's equivalent resolution 285
Shades of gray 194
Shield 58
Shot noise 108
Shrinking raster 179
Shutter speed 24
Signal-to-noise ratio 119, 160, 295
Size, video chip 86
Sky, ,night 36
Sky-to-ground ratio 315
Smear 56, 60, 64, 68
SNR_{TH} 310
Source, calibration 32
Spatial frequency 235
Spectral response 93
Spurious resolution 252
Square-root integral 289

Square-wave response 229
Streaking 70
Strobe light 64
Sub-array 56
Subcarrier frequency 134
Subjective quality factor 288
Summary noise factor 307
Super pixeling 53, 258
Superposition 225, 228
Synchronous integrator 300

Target transfer probability function 318
TDI 66, 146, 245
Temperature
    color 19, 34
Three-dimensional noise 301
Time-delay and integration
    see TDI
Transmittance
    atmospheric 315
TV limiting resolution 280
TV lines 236
Two-dimensional MRC 313

Undersampling 200

Velocity error 245
Vertical overflow drain 78
Video
    component 131
    composite 131
    EIA 170 125
    EIA 170A 125
    EIA 343A 130
    NTSC 125
Video cassette recorder 8
VIDEO FORMAT 125
VIDEO TIMING 127
Viewing ratio 172
Virtual phase 52
VISIBILITY CODE 316
Visual acuity 172

Well capacity 43
White noise 110
Wien's law 19

X-ray transfer 90